PLEASED to MEET ME

PLEASED to MEET ME

HOW GENES, GERMS, AND THE ENVIRONMENT MAKE US WHO WE ARE

BILL SULLIVAN

Washington, D.C.

Published by National Geographic Partners, LLC
1145 17th Street NW Washington, DC 20036

ISBN: 978-1-4262-2055-5

National Geographic Partners
1145 17th Street NW
Washington, DC 20036-4688 USA

Interior design: Nicole Miller

Printed in the United States of America

To my kids, Colin and Sophia.
There's a lot of me that I see in you.
At least you got some good traits
from your mother.

CONTENTS

» INTRODUCTION «

MEET THE REAL YOU

People do the strangest things, don't they?

But no matter how normal you think you are, there are people out there who think *you're* the oddball. From our diet to our habits to our beliefs, humanity is gloriously diverse.

How does this play out? Well, some people enjoy exotic foods and fine wine; others want nothing more than a plain hamburger and a Bud Light. Some people are vegetarians whereas others say that brussels sprouts taste like fart truffles. Some people stay thin throughout their lives; others feel their thighs widening just thinking about cheesecake. Some like to work out, and others would rather chill out.

We are collectively diverse in our habits as well. Some people wear sports jerseys and war paint to root for the home team; others would rather cosplay as a Borg at a Star Trek convention. Some live for a wild night on the town, and others prefer a night at the museum. Some people like to globe-trot; others won't even venture out to World Market. Some people are fashionable, and others would wear down the hosts of *What Not to Wear*.

And what about our behavior? Alcohol and drugs have no pull over some, but others cannot escape their gravity. Some people are always honest, yet others will lie, cheat, and steal with no remorse. Some people are color-blind; others only want to see white. Some wouldn't hurt a fly, while others fly off the handle. Some fight for war, others fight for peace.

The same is true of our romantic inclinations. Some people are faithful to their partners; others pretend to be. Some bank on good looks and money, while others are more invested in what's below the surface. Some people want a soul mate to love for the rest of time; others would view that as a life sentence. Some people remember anniversaries; some people are forgetful.

And what of our very natures? Some people are kind, and others are mean. Some have boundless energy, while some seem lazy. Some are fearless, and some are scared of their own shadow. Some always see the glass as half full; others always see it as half empty and leaking.

And between all of these extremes are many people who fall somewhere in the middle. We are all flesh and blood—but what tremendous variability in how we live our lives! Nevertheless, I trust we have one thing in common: a desire to understand why each of us is so spectacularly different.

THROUGH THE AGES, people have watched as philosophers, theologians, self-help gurus, and Frasier Crane have attempted to tackle the mysteries of human behavior—often with limited success. But practical answers to the questions of why we are who we are and do the things we do are coming from an unexpected source: research laboratories.

Scientists have recently learned a great deal about us: deep, dark secrets that everyone needs to know. The better you know your true self, the easier it is to navigate life's journey. And, by knowing what makes people tick, you will have a better understanding of those who are not like you.

All of us like to think that we march to the beat of our own drum. But science has revealed that the rhythm is played by percussionists we can't see with the naked eye. We march through life believing that we're the drummer—but the shocking evidence reveals this is an illusion. The truth is that there are hidden forces orchestrating our each and every move.

To illustrate this point, let's consider one of my personal quirks: my distaste for vegetables like broccoli. I've always hated broccoli because it tastes so bitter to me; the smell of it cooking can make me gag. My wife, however, eats a lot of it. Willingly! What's the difference between us? A clue comes from how our children responded to broccoli as infants: My son liked it, but my daughter reacted as if we were trying to poison her. We did not teach our children to love or loathe broccoli; they came to us this way, suggesting that this behavior is written in our DNA (we'll spell out how this works in Chapter 2).

Let the ramifications of that sink in a minute: The genes in our DNA have a say in whether or not we like something. I'm vindicated! My aversion to broccoli is not my fault, and I should stop apologizing for it because I had no say in which genes I acquired.

If we're not in control of something as basic as our personal tastes, what other things about us are beyond our command? In these pages, I set out on a quest to see how much genes contribute to our behavior. As we'll see, DNA presides over much more than our physical attributes like eye color and whether we're born with hands. It can also influence what we do with our lives, how quickly we lose our temper, whether we crave alcohol, how much we eat, who we fall for, and whether we like to jump out of perfectly good airplanes.

DNA is often referred to as the "blueprint of life" because it contains the instructions to build an organism. When it comes to building most people, DNA engineers the biological equivalent of a humble abode—although some folks get a mansion while others receive what we'll call a fixer-upper. And then some seem to be constructed from the blueprints of the Death Star.

But surely, we are more than a pile of genes, right? Your relatives, for example, share a lot of your DNA, but you can all be surprisingly different. Even identical twins, who are essentially genetic clones that share 100 percent of their genes, often diverge in their appearance and behavior. The home-remodeling TV show *Property Brothers* is hosted by identical twins, but they are not exactly the same. One is a quarter inch taller than the other. One is obsessed with fashion and enjoys wearing suits, while the other dresses casually. One enjoys hammering out the business details; the other prefers to swing a hammer. One is a conscious eater while the other has a relaxed approach to diet. These differences suggest that genes build a house but something else makes it a home. In this book, we'll take a look at the factors in our environment that can affect how our genes work, as well as how the environment may alter our DNA in ways that might be passed on to future generations. The means by which the outside world interplays with our genes is a new field of study known as epigenetics.

Epigenetics can have a tremendous impact on our behavior—and, remarkably, its effects on our DNA begin *before we're born*. For example, exposure to nicotine or other drugs may chemically alter genes in the sperm of a father-to-be. What a mother does during pregnancy can also introduce lifelong changes onto the baby's DNA. Epigenetics may play a wide-ranging role in obesity, depression, anxiety, intellectual ability, and more. Scientists are discovering how stress, abuse, poverty, and neglect can scar a victim's DNA and adversely affect behavior for multiple generations. These astonishing findings in epigenetics constitute another hidden force directing our behavior, over which we have also had zero control.

In addition to our own genes, scientists have recently recognized that microscopic invaders bring a massive repository of genes into our bodies that are likely to shape our behavior as well. Ever hear of a microbiome? Well, pull up a stool and pay attention, because we're going to learn all about it. The first microbial stowaways to set up camp in our gut came from our mother. We pick up more microbes from our food, pets, and other people as we age. New studies are revealing that the trillions of microbes teeming inside our guts may exert an influence on our food cravings, mood, personality, and more. For example, scientists can make a normally perky mouse gloomy by replacing its intestinal bacteria with microbes taken from a person suffering from depression. We'll examine how the Western diet many of us enjoy radically changes the composition of gut bacteria, leading some to speculate that it could be a contributing factor to health problems like allergies, depression, and irritable bowel syndrome, which are more common in affluent countries.

There is also a one in four chance that a common parasite transmitted by cats—the parasite that we study in my laboratory—might have hijacked your brain, dampening your cognitive abilities and predisposing you to addiction, rage disorder, and neuroticism.

We'll discuss emerging evidence suggesting that all of these tiny microbes are affecting our behavior for their benefit, making us wonder yet again if we are really in full control of our actions.

WORKING IN THE BIOLOGICAL SCIENCES for the past 25 years has provided me with a unique perspective on how life truly operates. My research into the hidden forces underlying our behavior has convinced me that almost everything we think we know about ourselves is wrong. And we are paying dearly for it. Our false sense of self hurts our personal, professional, and social lives. Our collective misunderstanding of human behavior impedes progress and adversely affects education, mental health, our justice system, and global politics. Exposing these hidden forces provides important new insights into our behavior, as well as a better understanding of people who do things we would never dream of doing.

In the upcoming chapters, we will take a closer look at just how much—or really, how little—control we have over our actions. This knowledge will help us better ourselves, and has the power to change our behavior in ways that will lead to a happier and healthier world. We'll review the biological reasons underlying obesity, depression, and addiction, and how that knowledge is paving the way toward better treatments for these conditions. We'll learn the real reasons why some people become aggressive or murderous, revealing potential ways to prevent these hideous acts from occurring. We'll also explore what science is teaching us about love and attraction, and how these lessons can improve our relationships. And finally, we'll peer into the psychology of our beliefs, including our political differences, in hopes that we can understand what leads us to act with blind faith, rather than insightful reason.

I can't wait to tell you about you! But before we dive into humanity's wildly diverse spread of behaviors, we'll need to understand the hidden forces working behind the scenes to animate us. So let's begin our journey by meeting our maker.

» CHAPTER ONE «

MEET YOUR MAKER

It's not an easy thing to meet your maker.

—Roy Batty, *Blade Runner*

Think back to the earliest grade you can remember and picture the bright, youthful faces of your past friends and classmates. Like empty pages thirsty for ink, futures had yet to be written and the possibilities seemed boundless. Optimistic clichés like "You can be whatever you want to be!" were part of your everyday value system.

Now, with the image of those sunny young faces in your mind, think about who these people have become. Some of your old chums have stellar careers doing what they love; others hate their menial work, and some can't seem to hold on to any job. Most went to college, but some were lucky to finish high school. Some still love their high school sweetheart, but others change spouses like toothbrushes. Some may have married a member of the same sex. Some still live in your hometown, some have ventured away, and a few may be homeless. Some still have six-pack abs, and others have packed on a keg. Some are helicopter parents while others neglect or abuse their children. Some are always bubbly and happy; others make Morrissey look content. Some became addicted to alcohol or drugs, or became pedophiles or even politicians. A few may be in jail.

Why did everyone turn out to be so different? Our peers grew up at the same time, in the same place, around the same people, and yet we are far from uniform in our behavior. Maybe you saw signs of unusual quirks in some at a very early age. Little Charlie loved sniffing glue. Kate's been sneaking candy since preschool. Young Cameron was rather effeminate, and Donald never cared about anyone but himself. And there just wasn't something right about creepy Carrie.

When we look at our peers who have been successful, many of us assume that they had gumption, determination, and a strong work ethic. Similarly, we are quick to blame those who haven't been as successful as being weak-minded, undisciplined, and lazy. If the story of your life reads like a Pulitzer Prize winner, then you deserve praise. If it reads like a cheap paperback better suited for lining a birdcage, then you deserve blame. Either way, most people believe that whether or not you are a success is all on you.

The idea that we're the masters of our own fate resonated with me while growing up. But as I learned more about biology, this simplistic concept was no longer edifying. Take overeating, for example. Many people blame the people with obesity and scoff that they have no self-control. But that really doesn't tell us anything useful, does it? *Why* do some people lack self-control? Same of individuals with depression. Some who don't know better dismiss the problem: "Put on your grown-up pants and get over it already!" Again, that doesn't help. *Why* can't people with depression snap out of it? Our rationale for murderers is equally impotent when we say, "Their soul is pure evil." Well, *why* were they moved to violence? We need to dig deeper if we have any hope of truly understanding our actions.

When our computer takes a long time to open a program, we don't think it's being lazy. When our car won't start, we don't yell at it for lacking determination. If a plane's engine fails and forces an emergency landing, we don't consider the plane to be guilty of malice. Granted, we are far more sophisticated machines, but machines we are. As Captain Jean-Luc Picard said of the humanlike android, Data, in *Star Trek: The Next Generation,* "If it feels awkward to be reminded that Data is a machine, just remember that we are merely a different variety of machine—in our case, electrochemical in nature."

The good captain and today's biologists do not say such things to dehumanize us, but to reveal what being human really means. If we understand how our biological machine works, then we are in a position to understand behavior and to fix it if need be. But we're like *The Greatest American Hero,* who had a red suit of superpowers but no instructions on how to use it. Understanding our behavior would be much easier if we had an owner's manual. And in 1952, scientists Alfred Hershey and Martha Chase found it.

In their hunt for the substance that holds the instructions to build an organism, Hershey and Chase looked to the simplest life-form they could find: a type of virus called phage, which infects bacteria. Made of only protein and DNA, phage viruses look like little Apollo moon landers that settle onto the surface of bacterial cells. Hershey and Chase tagged each component of the phage separately using radioactive atoms. They used radioactive phosphorus to label DNA and radioactive sulfur to label protein (there are no sulfur atoms in DNA and no phosphorus atoms in proteins). By following the different radioactive atoms, they could detect where the phage DNA and protein were before and after the phage infected the bacteria.

As it turned out, the phage DNA was injected inside the bacteria while its protein shell remained outside on the surface. Once inside, the phage DNA specified the building of more phages, until so many were built that the bacterium exploded. This elegant experiment showed that DNA contains the instructions to build baby phages (or any kind of baby, for that matter).

DNA has the shape of a double helix, resembling a spiral staircase in which each step is composed of a pair of biochemicals called nucleic acids (there are four nucleic acids in DNA, abbreviated A, T, C, and G). Such a structure makes it easy to see how DNA carries the units of heredity we call genes. The spiral staircase can unwind to look like a ladder, and the two nucleic acids making up each rung can be separated like pulling down a zipper. When DNA's fly is down, its code is exposed and is "transcribed" onto a carrier molecule called messenger RNA (mRNA) to make a protein. If we consider DNA the foreman, proteins operate like construction workers, providing our cells and tissues with their structure and function.

Hershey and Chase's work suggested that DNA contains the genes needed to build an exact replica of an organism, a clone. This theory became a reality in 1996, when Dolly the sheep was born: the first mammal cloned from an adult cell. Dolly was created by putting DNA from an adult sheep cell into an egg that had its DNA removed; that egg was then implanted into a surrogate mother. Dolly was named after Dolly Parton, because the adult cell's DNA used for making Dolly came from the progenitor sheep's breasts (I'm not making that up!). Using the same technique, the first monkeys were cloned in 2018.

In 2003, the Human Genome Project completed sequencing the three billion nucleic acids that comprise human DNA. That is a lot of information—stretched out, the DNA from just one of our cells is two meters, about the length of a queen-size bed. If we read our DNA sequence one letter a second, it would take nearly a hundred years to finish. Our genome contains about 21,000 genes scattered across 46 chromosomes, 23 from our mother and 23 from our father.

DNA has been toiling away for eons, creating all sorts of different life-forms suited to various environments. Life has been in the reception room for at least 3.5 billion years. But now one of its many creations has finally been called back to see the boss: We're the first species on the planet to meet our maker.

Why You Can't Be Whatever You Want to Be

Learning to read the language of DNA has forced us to rewrite our history book. The abundance of diverse life on Earth was not conjured up out of the blue, all at once. It started as a simple, single cell housing DNA, and evolved from there over the course of billions of years. Life-forms began to compete for resources, and those with characteristics that allowed them to thrive in their surroundings passed their DNA to the new generation, much as a relay runner passes a baton. Those that couldn't compete either died off or moved away and developed on a different evolutionary trajectory tailored for survival in their new environment.

Renowned biologist Richard Dawkins described genes as "selfish" replicators: the Gordon Gekkos of the biological world. He refers to the organisms that selfish genes build as "survival machines," because their fundamental purpose is to protect their DNA and ensure its passage into the next generation. Author Samuel Butler actually called it a century earlier when he said, "A hen is only an egg's way of making another egg."

Despite our fancy bells and whistles, we are no different. Scientists who study evolutionary psychology argue that virtually all of our behavior is motivated in some way by the single-minded drive to find a mate(s) and reproduce our genes. Through this lens, much of human folly comes into focus. The pull toward one-upmanship, greed, and power is but an undertow in our gene pool that many fail to resist.

Differences between people arise from differences in their DNA sequence. Although many acknowledge that DNA builds their fleshy edifice, most don't realize that genes have an impact on more complicated characteristics as well, such as intelligence, happiness, or aggression.

In some cases, how genetics affects our body is straightforward. Sometimes a change in a single gene, referred to as a mutation or variant, produces an alteration that is highly predictable. One example is sickle-cell anemia, which results when red blood cells are misshapen. It is caused by a mutation in the gene that makes hemoglobin, the protein that carries oxygen in red blood cells. People born with this mutation in their hemoglobin gene will develop sickle-cell anemia, no question about it.

In contrast, more complicated traits, such as those that influence personality and behavior, arise from many different genes working in concert with one another. Variations in a single gene within such a network don't always guarantee a perceptible change in the organism. This is why it is essential to keep in mind that most genetic variants tell us about *predispositions,* not certainties.

Think of our genes as the bricks in a Jenga tower. Pull out the wrong brick and the tower comes tumbling down. But pull out a different brick and the tower remains standing. As long as the other bricks can support the structure, we're still in the game. Similarly, a mutation in one gene does not necessarily spell disaster for our body; whether it makes us crumble depends on the other genes that support the mutated one. We should also keep in mind that not all gene variants are detrimental; like the X-Men, sometimes mutant genes grant us a superpower.

Despite the caveats, our genes can provide valuable insights into what we can be and what we cannot be. Here are some of the things I would like to be. I would like to sing like Steve Perry of Journey. I would like to be taller. It would be a refreshing change of pace to have women swoon when I walk by. Being smarter than Albert Einstein would be cool. I think it would be badass to have wings and fly like the Hawk Men in *Flash Gordon.* But try as I might, I will never be a tall stud who uses his own wings to fly to Stockholm to collect my Nobel Prize while singing "Don't Stop Believin' " as the finale of my acceptance speech. It's fun to dream, but we need to accept the truth: We can't be whatever we want to be. The genes we inherit at conception are like the playing cards we hold at the poker table: We have to play our best game with the hand we're dealt.

As Lady Gaga professed, we are "born this way," constrained by certain limitations that start at the genetic level.

And, as we'll soon see, DNA is just one link in the leash that pulls us through life.

How Your Environment Affects Your Genes

Imagine we made an identical copy of you, using the same method scientists used to produce Dolly the sheep. By inserting your DNA into an egg that had its DNA plucked out, we could implant a new you into a surrogate mother. Forty weeks later, she'll have a baby who looks exactly like you did. That baby will grow and be your spitting image every step of the way. But here's the million-dollar question: To what degree will your clone *behave* like you?

Sequencing the human genome was a giant leap forward in understanding how we function, but it only provides a rough sketch of the portrait of you. Your DNA sequence is not read like a typical novel, but more like a Choose Your Own Adventure book where the environment steers how the tale unfolds. Your DNA houses many different potential versions of you. The person you see in the mirror is just one of them, fished out by the unique things you've been exposed to since conception.

Your environment dictates whether a variation in your DNA becomes relevant. If I were born 50,000 years ago, I probably wouldn't have lived very long. It's not just that I loathe camping and barely have enough upper-body strength to open a bag of chips, but also that my nearsighted eyes would have made me a pathetic hunter-gatherer and easy prey for lions, tigers, and bears. Natural selection kept people with lousy vision out of the gene pool for eons. But with the invention of glasses, folks like me were back in the game.

The environment can have a direct effect on the fidelity of your genes. Random genetic mutations can arise, say from frying in the sun or falling into a vat of spent nuclear fuel. Radiation and certain chemicals are called mutagens because they can damage DNA, which often leads to cells gone wild: cancer. The number of potential mutagens rivals the number of albums sold by Taylor Swift. But some of the more familiar include ultraviolet light, tobacco, alcohol, asbestos, coal, engine exhaust, air pollution, and processed meats. The degree of exposure, combined with your genetic predispositions, determines the amount of DNA damage your cells may incur.

The environment can clearly change gene function by damaging DNA, but that is not the only way it influences how genes work. To better understand the next part, it may be helpful to think of your genes as the keys on a piano. If you tickle the ivories at random, it would sound like the music in a horror flick. The right keys must be played at the right time to create a beautiful song. Your genes must work the same way. If they were played all at once, you'd look like Freddy Krueger.

Every cell in your body contains the same 21,000 genes, so how are some brain cells and others butt cells? Only brain cell genes are switched on (expressed) in your brain cells. The genes for butt cells are still present in brain cells' DNA; they just aren't expressed (except maybe in that butthead ex of yours). Proteins called transcription factors control whether a gene is expressed by binding to a DNA sequence called a promoter, which lies at the start of the gene. Transcription factors govern whether a gene is turned on or off, acting as activators or repressors, respectively. When you were an embryo, you were composed of stem cells that had the potential to become any cell type in your body. The transcription factors in that cell largely dictated the fate of your embryonic stem cells. Transcription factors that activated brain genes were present in the stem cells that became your brain. Those that activated butt genes were in the stem cells that became your butt.

Many things, such as hormones, influence the activity of transcription factors. Made by your endocrine system, hormones control development, sex drive, mood, metabolism, and more. Many substances in the environment act as endocrine disruptors, meaning they mimic a hormone's activity and disrupt gene expression accordingly. Consequently, endocrine disruptors may cause developmental, reproductive, neurological, and immune defects. Endocrine disruptors include certain medications, certain pesticides, and the bisphenol A (BPA) used in plastics. As with mutagens, the amount of the endocrine disruptor dictates whether it has a significant effect on the activity of your genes. The jury is still out on how much is too much, but it is important research because endocrine disruptors are everywhere (including in many items that pregnant/nursing women and children use). In addition, the negative effects of endocrine disruptors may affect children for multiple generations. A 2018 study reported that a mother's exposure to the endocrine disruptor diethylstilbestrol (DES) increased the risk of attention deficit/hyperactivity disorder (ADHD) in her grandchildren.

Transcription factors are central to regulating gene activity, but they do not work in isolation. As scientists began to study DNA more closely, it became evident that it is not a uniform molecule. Some sections of DNA are tightly wound up and compacted, whereas other sections are relaxed and open. Genes in compacted DNA are not expressed as much as genes in the open sections. Cells can control a transcription factor's access to genes in DNA in two major ways. The first is DNA methylation, which occurs when a small chemical called a methyl group is directly attached to the nucleotides that make up a gene. With methyl groups strewn across a gene, it becomes harder to read, much like someone blacking out the letters in a sentence. As such, a methylated gene moves toward the "off" position, or is silenced. A second mechanism involves a group of proteins called histones, which form spools that DNA wraps around like a piece of thread. Histone proteins are subject to numerous chemical modifications that affect expression of the gene associated with them. With transcription factors, these processes provide an incredible amount of flexibility for gene expression, allowing genes to be fine-tuned rather than just on or off. It is more accurate to think of gene expression as a dimmer switch, rather than a light switch.

Processes that affect gene expression without changing the DNA sequence itself are "epigenetic," meaning "beyond the gene." Epigenetic modifications (also called epigenetic marks) allow the environment to send a text message to your genes that not only alters how they work for you, but also how they may work in your kids and grandkids. As the famous botanist Luther Burbank noted, "Heredity is nothing but stored environment." Physical substances that you encounter in the environment can produce epigenetic changes to your DNA, altering which genes in your body are being expressed. This can be a great advantage to you and your children, because the rapid changes in gene expression allow quick adaptation to environmental conditions.

Remarkably, in addition to physical substances altering gene expression through epigenetics, certain behaviors like child abuse, bullying, addiction, and stress can do so as well. Negative events can scar our DNA, and in certain cases, these scars are passed along to our children. We will see several examples of this in chapters to come, but here's one that illustrates the importance of epigenetics on our behavior. It is well established that low socioeconomic position correlates with increased disease in adulthood; children raised in poverty are much more likely to be unhealthy as adults. This may be so for many environmental reasons, but some differences right out of the starting gate could be extremely important, too. In a 2012 study, geneticist Moshe Szyf at McGill University in Canada showed that different groups of genes are methylated in adults who suffered economic challenges in early childhood compared to those who were well off. Similar differences in DNA methylation are seen in monkeys born into low rank compared with those born into high rank.

These studies, and many more we'll discuss, suggest that our DNA is front-loaded with epigenetic marks received in early childhood or while we're still in the womb, the latter of which is called fetal programming. Could we be born preprogrammed to behave according to where our genes think we lie in the social hierarchy? Could these differentially methylated genes in impoverished youth help explain health or behavioral problems later in life, locking families into a vicious cycle? We don't know the answers to these provocative questions yet, but studies like these suggest that not only are underprivileged children suffering adverse social conditions, but they also suffer adverse biological consequences.

The epigenetic marks added to our histone proteins early in life could also affect our behavior. Epigenetics may even dictate our career decisions, particularly if we're ants in the laboratory of biologist Shelley Berger at the University of Pennsylvania. Members of an ant colony carry out specialized tasks; larger major ants are soldiers that defend the colony, while smaller minor ants are foragers that collect food for the colony. You might think that majors enlisted in the ant military, while minors learned how to scavenge from experts, but that's not how it works.

Because these behaviors are not taught, Berger and her colleagues hypothesized that epigenetic mechanisms swayed the ants' lot in life. To test this, she injected a drug into the brain of baby ants that altered the histone proteins interacting with DNA. The first surprising thing is that it's possible to inject something into a baby ant's brain. Second, by altering the histones, Berger was able to reprogram an ant's behavior, turning a soldier into a forager (foragers who got the drug foraged even more than usual). In other words, this epigenetic drug changed the soldier ant's destiny without changing its genes.

The study of epigenetics emphasizes the intimate interaction between our genes and our environment, and reveals why our genes are not necessarily destiny. Although we had no say in which genes we were dealt at birth, we may be able to alter our environment in ways that affect how those genes are expressed, much like an expert poker player might bluff her way to winning a game with a junk hand.

How Microbes Add to Your Genes

Scientists have recently appreciated that more than the 21,000 genes in our DNA affect our body. We have trillions of microbes—bacteria, fungi, viruses, and parasites—living on and within us that contribute *millions* of additional genes to our genetic ecosystem. That might give you the willies, but the vast majority of these tiny stowaways, collectively referred to as our microbiota (and their genes our microbiome), come in peace and bearing gifts. For example, bacteria in your gut help you digest food and make vitamins. Some sulfur-producing bacteria can give you the power to clear the room when you'd rather be alone. These "friendly" bacteria, which do not cause disease, also help keep "unfriendly" pathogenic bacteria in check.

We have our mothers to thank for starting our microbiota collection. We are coated with our first bacteria as we glide through the birth canal. Our mother continues to share her bacteria with us through breast-feeding. The microbiota are therefore somewhat heritable, because some species are transferred from mother to child. We continue to acquire microbes throughout our lives, picking them up from food, water, air, doorknobs, and interactions with other people and animals. People around the world have different types of bacteria in their gut depending on such things as diet, geography, hygienic standards, illness, and age.

You've probably noticed that everyone's home smells a little different. Sometimes this is due to cooking, pets, smoking, mildew, or teenage boys—but it is also due to the microbiomes of the inhabitants. Researchers have found that, like Pig-Pen from Peanuts, you are surrounded by a "germ cloud." You leave pieces of your microbiota wherever you go, like a trail of microscopic breadcrumbs.

Armed with this information, it might even be possible in the not-too-distant future for police to use microbiota to track people, the way they currently use fingerprints or DNA. Our germ cloud likely contributes to how dogs can track people so easily, and why mosquitoes bite some of us more often than others. The by-products generated by the bacteria living on our skin produce a scent that is released into the air as we move. Animals with a keen sense of smell can get a whiff of these aromatic compounds and follow them to the source. As we'll see in Chapter 7, our germ cloud may also influence with whom we'll have a stormy romance.

These microbes are tiny, but as Yoda cautioned, we should not judge things by their size. About 10,000 species of bacteria reside in our gut, supplying us with an extra eight million genes. Their collective weight is up to three pounds, which means our microbiota weigh more than our brain. It's also good news if we're on a diet. As you stand on the scale tonight, feel free to apply this new knowledge and subtract three pounds of bacterial weight from your body weight. (You're welcome!) And here's another microbiota factoid you can use to bewilder guests at your next party: Bacterial cells in our body outnumber human cells, meaning that we are more bacterial than human. With so many other creatures living on and inside us, just how much are they running the show?

In recent years, the microbiome has received a huge amount of press. The microscopic creatures in our body seem to exert an influence on just about everything, from appetite to wound healing. In addition to producing vitamins and other dietary compounds useful to our bodies, gut bacteria are a major source of neurotransmitters, the biochemicals that act on our brain. Some scientists have suggested that by virtue of producing neurotransmitters, our bacteria may modulate our moods, personality, and temperament.

When researchers raise mice that are deprived of their microbiota, the mice exhibit strange neurological problems and do not respond to stress properly. These studies exposed the gut-brain axis, a conduit of biochemical communication between these organ systems. Such an axis exists in people too, as researchers have noted a strong correlation between intestinal problems and mental illness. For example, anxiety and depressive disorders are strongly associated with irritable bowel syndrome and ulcerative colitis. In addition, parasites that do not kill are present in many people; these parasites can sit dormant in the brain for the rest of their lives. As we'll discuss, scientists have correlated the presence of a common parasite in three billion people with certain behaviors.

Through the genes they bring into our body, our microbial inhabitants constitute another hidden force that tugs on our behavior strings in ways that are completely unbeknownst to us.

Why Our Maker Is in Trouble

In the *Star Wars* films, Sheev Palpatine (the Emperor) became the master of Darth Vader after turning him to the dark side. But in the end, Darth Vader destroyed the Emperor. It's a classic tale in which the servant kills his master. A similar fate may await genes, which have been the undisputed masters of the Earth for nearly four billion years.

Roughly 600 million years ago, genes constructed the first neuron (brain cell) in ancestral organisms that may have resembled modern jellyfish or worms. In the many years since, these neurons have banded together to form brains, giving the lucky survival machines that carried them a new advantage. Over time, brains became larger and faster as they amassed more neurons and increased the number of connections between them. In addition to humans, the brains of some animals have become powerful enough to reach self-awareness (including nonhuman primates, elephants, dolphins, orcas, and magpies). The evolutionary path that developed the brain was a yellow brick road, and it led us to the discovery that DNA is the wizard behind the curtain.

Our brain imparts a sense of self that makes us feel like the Decider, and it's tempting to believe that it frees us from the tyranny of genes. One limitation to this enticing idea is the inescapable fact that our thinking organ was constructed from the genetic blueprint in our DNA: The brain is an organ of our genes, by our genes, and for our genes. But as we'll see, brains are not created equally, and we did not get to choose the one that ended up between our ears.

Despite its initial genetic constraints, does the brain develop enough sophistication to take on a life of its own, to think for itself? Our brain is composed of a mind-blowing 100 billion neurons, which is 1,000 times the number of people following Katy Perry on Twitter. Moreover, on average, a single neuron can have a staggering 10,000 projections connecting to other neurons, allowing them all to chat with one another using biochemical signals. A human brain has more than 100 trillion neural connections, which means we have 1,000 times more brain cell connections in our head alone than stars in the Milky Way.

As in other animals, most of our body's physical operations such as heartbeat, breathing, digestion, and sweating are on autopilot, controlled by the oldest part of our brain. Sitting on top of this automated system like swirls of soft-serve yogurt is our large cortex, the part of the brain that contemplates the weather, stock market, what just happened on *Stranger Things,* and whether you should accept that friend request from your ex.

This massive chat room of neurons brings the outside world into our head and debates how to respond. And the plot thickens. As the control center of a highly social species, our brain works in the context of countless other brains: a gargantuan hive mind of information from past and present. Now that our collective brains have caught on to DNA's selfish game, what will be our response?

We will soon have the power to give our maker a makeover. We are developing ways to edit genes, manipulate epigenetic marks, remodel microbiomes, and modulate the brain—activities that are making us co-authors of life, rather than just passive readers. Our ability to create self-replicating machines with artificial intelligence may dispense with the need for genes altogether. Will we merge biological life with mechanical life, or are we merely stepping-stones in a universe destined to be occupied by androids? If we're not careful, we may share Darth Vader's fate of conquering our DNA master but being mortally wounded in the process.

Science is revealing a great deal about who we are and why we do the things we do. But our owner's manual is more complex than we ever imagined. Despite our intelligence, humor, and love of the arts, we must acknowledge the core of what we are: a survival machine built by DNA that lives under the influence of numerous hidden forces that are beyond our control. In the upcoming chapters, we will take a closer look at just how much, or how little, control we really have over our actions—and how we can use this knowledge to better ourselves and others in this world we share.

» CHAPTER TWO «

MEET YOUR TASTES

I do not like broccoli. And I haven't liked it since
I was a little kid and my mother made me eat it.
And I'm President of the United States
and I'm not going to eat any more broccoli.

—George H. W. Bush

There is no question that broccoli is good for you—but for me and about 25 percent of the population, it tastes like dog's breath. Ditto for kale, brussels sprouts, cauliflower, and most of the other cruciferous vegetables parents mercilessly force upon our protesting taste buds. My aversion to these popular veggies has made me the object of ridicule at many social dinners. There are so many fascinating topics that we could talk about, but the conversation inevitably turns into another annoying inquisition probing my dietary habits.

"You don't even like salad?!" No. If a plate of salad is in front of me, I react just as Ron Swanson did on *Parks and*

Recreation: "There's been a mistake. You've accidentally given me the food that my food eats."

"It's probably a psychological trauma. Did your mom shove broccoli down your throat as a child?" No. I would stuff the entire portion into my mouth and claim that I had to use the bathroom.

"You're a scientist, surely you realize how nutritious vegetables are?" Yes, but for this scientist at least, it's not easy eating greens. I'll take the carrots instead.

Sometimes I feel like I need to have surgery to remove these people from my back, but when I'm the one getting grilled at a cookout, I can't help but wonder what is wrong with me. I see someone shovel in a mouthful of some green vegetable—deliberately—and then genuinely enjoy it. I am green with envy.

Certain vegetables aren't the only item on the menu that might cause contention among diners. Some people have a major sweet tooth. Some love spicy foods. Some can't handle dairy products. Some can't function without their coffee. Some don't like to drink alcohol, and others are very finicky about their wine. And some people like to eat bizarre items that many would not consider edible. Everyone's tongue looks the same, so why do our tastes in food and beverage vary so much? Is there hope of achieving peace at the dinner table?

Why You Hate Broccoli

Our varying fondness for broccoli was famously played out in the "Chicken Roaster" episode of *Seinfeld*. Kramer is protesting the Kenny Rogers Roasters restaurant, but becomes addicted to their food after trying it. He then devises a covert operation to have his buddy Newman buy meals from the restaurant for him. Jerry gets suspicious when Newman is caught buying broccoli at the restaurant, because Newman "wouldn't eat broccoli if it was deep-fried in chocolate sauce." To dispel Jerry's suspicions, Newman claims to love broccoli. But when Jerry challenges Newman to eat a piece, he quickly spits it out, calls it a "vile weed," and does a honey mustard shot to dampen the bitter taste.

Newman is clearly a "supertaster," a term physiological psychologist Linda Bartoshuk coined to describe people like me who have a heightened sense of taste. To be a supertaster might sound like a good thing, but it's not. Instead of an "S" on my chest, it's more like a scarlet letter on my forehead.

Do you think you might be a supertaster, too? Well, you can test yourself. A bit of blue food coloring on your tongue will stain everything except the taste buds, which will appear as pinkish bumps. Stick one of those loose-leaf paper reinforcement circles onto the tip of your tongue and tally up the taste buds in the circle using a magnifying glass. Supertasters tend to have more taste buds, usually 30 or more, within that circle.

Each taste bud is made up of about 50 to 150 taste receptor cells. A gene family called TAS2R (appropriately pronounced as "taster") makes the taste receptors on these cells that bind to molecules in our food or drink. After these molecules get into our mouth and bind to our taste receptors, the signal is relayed to our brain. *Ahhhhhh, Reese's Peanut Butter Cups* . . . or, *Oh crap, kale!*

In addition to having more taste buds, supertasters can also have genetic variations in their TAS2R genes that make their taste receptors better at detecting bitter flavors. A TAS2R gene called TAS2R38 registers thiourea compounds present in many vegetables. Hard to imagine that even a vegetarian's diet contains something as sinister sounding as thiourea, but this is just one of the many chemicals that make up broccoli. This is why scientists recoil at the self-proclaimed "Food Babe" Vani Deva Hari, who once warned, "There is just no acceptable level of any chemical to ingest, ever." All of our food is made of chemicals, even if it is organic and non-GMO.

In the 1930s, Arthur Fox, a DuPont chemist, was the first to note the different reactions people have to thiourea compounds. Fox accidentally splashed some of these chemicals onto himself and a lab mate; the chemicals didn't bother Fox, but his fellow chemist complained about their bitter taste. Fox was not a supertaster. His lab mate was. This was some of the first direct evidence that what one person tastes is not necessarily the same as what another person does.

The variations in TAS2R38 among people are due to differences in their DNA sequence at this gene, which essentially means the taste bud protein produced by that gene is going to be different. Specifically, the DNA of supertasters builds taste receptors that register thiourea compounds as incredibly bitter. A supertaster's brain assumes that the green horror he just stuck in his mouth must be unfit for human consumption. Now, broccoli won't actually make supertasters physically ill. But the bitterness is so potent that it can sometimes trigger their gag reflex. Said another way, the TAS2R38 variation in supertasters is DNA's attempt to play it safe and protect them from potentially poisonous plants.

It is important to remind ourselves that we are products of our DNA, the molecule that is single-mindedly dedicated to the mission of copying itself. DNA builds living creatures like us to serve as its survival machine and to maximize its chances of getting passed into another generation. (Sounds cold, but we're keeping it real here.)

As survival machines, we are equipped with taste buds to help us discriminate what might be useful to our bodies from what might be lethal. To understand our tastes, we need to recognize that plants are survival machines too. Because plants can't flee predators, their DNA has developed alternative protection strategies. One tactic is to make their parts unpalatable or downright toxic so that animals will stop munching on them. By producing bitter-tasting chemicals, plants can deter broccoli haters like me from making them lunch.

One strategy plants use for reproduction takes advantage of animals that have a sweet tooth. Such plants surround their seeds in a sugary fruit so animals will eat them and unwittingly spread the plant's seeds around. Plants are very manipulative when you think about it. If I could eat salad, I would eat it angrily, stabbing my fork through those hearts of romaine with gusto.

Why You Love Broccoli

If TAS2R38 variations protect us from eating poisonous plants, why don't we all hate broccoli? It most likely depends on the types of plants that were present in the environment of our distant ancestors. If our ancestors evolved in an area filled with poisonous plants, having the supertaster gene could have conferred a survival advantage. On the other hand, this blessing could become a curse if those plants are indeed edible; in this case, the supertasters cannot reap the nutritional benefits, because their taste buds have misled them.

Many other genes besides taste bud receptors influence what flavors we find palatable and how we metabolize (or break down) certain foods. Finding and characterizing these genes is a new science called nutrigenetics. In a 2016 study, geneticist Paolo Gasparini at the University of Trieste in Italy uncovered 15 new genes linked to people's preferences for various foods—from artichokes to yogurt. He identified these new genes by combing through the genome sequences of more than 4,500 individuals to find genes linked to 20 different foods these people liked. Interestingly, none of these genes are the usual suspects of smell and taste receptors, meaning we still have a lot to learn about why our bodies give a thumbs-down to certain foods.

Why You Can't Say No to Sugar

Not much can happen during your day that a little chocolate can't fix. But, believe it or not, not all mammals share a love for sweets. Did you ever try to break off a piece of that Kit Kat bar for your feline friend? Wonder why your gracious act was met with icy indifference? Strict carnivores like cats don't have the taste receptors to detect sweetness. (Surely this explains Grumpy Cat?)

In our modern world, our taste receptors for sweetness actually get us into dietary trouble. In the old days, our primate ancestors relied on ripened fruits to supply their bodies with caloric energy. Because fruits contain the most sugar when ripe, we evolved a sweet tooth to make sure we get the best bang for our buck when extracting energy from food. Therefore, our love of sweets is deeply rooted into our evolutionary heritage and is a very hard habit to break. However, you may have noticed that some people will easily surrender a doughnut, while others will fight to the death for it.

A gene variant for a sweet tooth has indeed been found—and not everyone has it. These mutants walk among us, turning down desserts and making the rest of us feel guilty. (I'm pretty sure my wife has the sweet tooth gene, though. When I ask her to split a cupcake with me, she gives me the bottom half.)

A 2008 study carried out by nutritional scientist Ahmed El-Sohemy at the University of Toronto identified a variant in a gene called SLCa2, which correlates with a tendency to take two lumps of sugar instead of one. SLCa2 encodes for a protein called GLUT2, which brings glucose sugar from our blood into our brain cells where it is broken down for energy. Researchers believe that this change in the GLUT2 receptor interferes with glucose sensing, and as a result, the body doesn't have a reliable measure of how much glucose is in the blood. You could have a full tank, but your glucose gauge says you are only half full. So you have a second piece of cake, blissfully unaware that you're already sweetened. Studies in mice support this idea: Mice bred to lack GLUT2 will keep on eating even after their brain is marinated in glucose. In people, SLCa2 gene variants correlate with an increased risk for type 2 diabetes.

Why You Love Junk Food

Do you still think that turning down junk food is strictly a matter of will? What if I told you that a predilection for junk food could have been programmed into your DNA before you were even born?

As it turns out, mothers who eat a "junk food" diet rich in sugar, salt, and fat give birth to children who have a seemingly inborn desire for junk food as well. In humans, we think this is because the kids grow up in a household that eats poorly. No one would dispute that possibility, but experiments on lab rats suggest more might be going on than first meets the eye. Chew on this: A 2007 study showed that rat pups born to mothers fed a junk food diet during pregnancy developed an increased preference for fatty, sugary, and salty foods. Rat pups born to mothers who ate a healthy diet during pregnancy did not want the junk food.

How might this happen? Because it is highly unlikely that the fetus accrued gene mutations because of a mother's junk food diet while in the womb, scientists suspect fetal programming took place: The mother's diet alters the DNA of the unborn child at the epigenetic level. In other words, the junk food did not change gene sequences; rather, it changed the expression levels of certain genes. It would be like the time Fergie sang the National Anthem at the 2018 NBA All-Star Game; the lyrics were the same, but the song was very different. So although it is not surprising that kids who grow up regularly consuming junk food are likely to become junk food junkies, numerous studies suggest that a proclivity toward junk food could have been programmed into the fabric of their DNA before the cord was cut.

One major way DNA can be epigenetically programmed is through methylation, a chemical modification to the DNA that affects a gene's expression. The more a gene is methylated, the less it is expressed. If you think of gene expression as a highway,

DNA methylation marks would be like a bunch of orange traffic cones strewn across that highway, slowing things down. A 2014 study took a look at the level of DNA methylation taking place at a gene called proopiomelanocortin (POMC) in the pups born to rats that were on junk food diets during pregnancy. The POMC gene gives rise to a key hormone that decreases appetite. Rat mothers who consumed a high-fat diet gave birth to pups that had higher levels of methylation at their POMC gene, which means that less of this appetite-suppressing hormone is made in these pups. So mothers gorging on junk food gave birth to offspring that were programmed in utero to be born hungrier than offspring whose mothers ate right.

What happens if the pups of junk food mothers were forced to eat a healthy diet? Is it possible to reverse the DNA programming that occurred in the womb? Unfortunately, this does not appear to be the case, at least in the aforementioned 2014 study: A healthy diet did not return DNA methylation levels at POMC to normal. In other words, the mother's junk food diet had permanent effects on a baby's DNA. If true in people, this might explain why it is so hard for some to control what or how much they eat. There may be a critical window during fetal development when DNA methylation is laid down in a permanent way.

Why You Think Cilantro Tastes Like Soap

Cilantro, the leaves of the coriander plant native to the eastern Mediterranean, is added to season a wide variety of foods, including salsa, seafood, and soups. Most people find the flavor delightful, but some spit it out, complaining that it tastes like soap. How they came to be an expert on "soap dishes," I do not know. But it is clear that some people despise cilantro. Even the famous chef Julia Child wasn't shy about her disdain for the herb, proclaiming that she would pick it out of her food and throw it on the floor.

Julia and her fellow cilantro haters are sensing chemicals called aldehydes in the herb, which are also found in—surprise!—soaps and lotions. So for them, cilantro literally smells like a bath product, and not a culinary seasoning. Smell and taste are intimately connected and, just as genes like TAS2R influence our taste receptors, genes also influence our odor receptors. A study of twins revealed a genetic component to one's fondness for cilantro. Identical twins were much more likely than fraternal twins to agree on their cilantro preference. Because identical twins share 100 percent of their DNA and fraternal twins share only 50 percent, the survey result points to a genetic component to our feelings about cilantro.

To hunt for the culprit genes, a research team at the genotyping company 23andMe surveyed 30,000 people and found that cilantro preference is linked to a gene called OR6A2. Consistent with what we know about the chemical composition of cilantro and soap, OR6A2 codes for an olfactory receptor that is highly sensitive to aldehydes. In a different study, cilantro preference was also linked to variants in three additional genes, this time including a TAS2R gene. Similar to what we saw for bitter foods, a genetic influence beyond our control is behind our ability to stomach certain herbs.

While we can't control the genes we were dealt at birth, there may be ways to neutralize the soapy essence emanating from cilantro. One way is by crushing the leaves to release enzymes that degrade the aldehydes. Or, if you have friends who are truly resistant to giving cilantro a chance, perhaps just accept them for who they are and make their dish with parsley instead.

Why You Like It Hot

As long as I can remember, my daughter has loved chips and spicy salsa. Even as a toddler, she would continue to shovel them into her mouth as tears streamed down her face. With reddened cheeks and smoke coming out her ears, she'd ask for more. She noted at dinner one night that her other favorite food, ketchup, was red like salsa. So why is salsa hot but ketchup not?

The answer to her question relates to how plants reproduce. Some plants have evolved ways to get specific animals to help them multiply. The hot peppers in salsa are from plants that use birds instead of land animals to disseminate their seeds far and wide. Peppers are loaded with noxious chemicals that make most animals feel like lightning struck their tongue, and yet birds don't feel the heat from eating peppers.

That's a clever strategy for plants, and most animals get the hint and leave the hot peppers for the birds. But not humans. We not only devour their flaming fruits with glee, but also breed plants to make peppers hotter than natural selection has ever aspired. The heat intensity is measured in Scoville heat units (SHU), named after the pharmacologist who devised the scale in 1912. For reference, a bell pepper has zero SHU and a jalapeño can have up to 10,000 SHU. The familiar tabasco pepper averages 40,000 SHU, and the hotter habanero chilies can reach 350,000 SHU. Some of the peppers purposefully bred to be the hottest on Earth, like the Carolina Reaper or Dragon's Breath, can reach upward of an incredible two million SHU. Pepper X, the first to break three million SHU, was debuted in 2018. These peppers nearly send the greenhouse up in flames.

People who have a high tolerance for spicy foods can thank the same genes that help us respond to hot temperatures. A gene called TRPV1 makes a type of protein receptor on the surface of our cells that physical heat activates. When heat melts a part of this receptor, it sends a message to our brain that says, Damn, this is hot! Spicy foods contain a chemical called capsaicin, which can also bind to those heat-activated TRPV1 receptors. When capsaicin activates TRPV1, it sends the brain the same message: Damn, this is hot! Our silly brain even thinks we're out in the heat, so it signals our eccrine glands to release sweat. We literally feel a burn because our brain gets the same message whether we are licking a curling iron or chomping on a ghost pepper. Alcohol activates TRPV1 too, which is why we feel that characteristic burn when we swallow a shot of whiskey.

Genetic variations in our TRPV1 receptor can weaken its ability to bind capsaicin, so the tolerance for spicy foods is going to be greater than for someone who has a version of TRPV1 that gives capsaicin a bear hug. Another reason why some like it hot is that they've developed a tolerance for capsaicin (just as people build a tolerance to alcohol or caffeine). In other words, these people might need more hot sauce than they used to just to feel that same level of heat they experienced as a sriracha virgin.

My dad exemplifies the connection between spicy food, physical heat, and tolerance. Not only does he carry plenty of hot sauce with him wherever he goes, but he will also ask that his already piping hot coffee be microwaved to one degree shy of spontaneous combustion. In support of the genetic connection, my dad's younger brother smothers his food in so much black pepper that you have no idea what he is eating—everything on his plate looks like a roof shingle. I clearly got my TRPV1 genes from my mom, as I usually need to wait for my coffee to cool down before taking a sip, my eyes water just staring at chili, and I prefer my Scotch on the rocks. My daughter's proclivity for spicy foods must have come from my wife, who buys buffalo sauce by the barrel.

For some people, though, the attraction to fiery fare does not appear to be related to tolerance. It's evident that some people, including my salsa-loving daughter, are indeed feeling the heat. They are crying, sweating, and howling in pain, yet continue to devour foods that taste like lava.

Why do such people relish the sensation of swallowing the sun? Studies have shown that many people who like spicy foods tend to be thrill seekers (which may foreshadow some tough times ahead for me as my daughter enters her teenage years). Enjoyment of spicy foods is referred to as benign masochism, which is a reasonably safe way to feel an adrenaline rush (kind of like watching a horror movie or disagreeing with someone's political beliefs on Facebook).

Of course, there are significant cultural explanations as well, as people who grew up on spicy foods gravitate toward them as adults. Cultures that tend to like things spicy typically live in hot climates, where food spoils quickly. Historically, people living in these regions made their food last longer by adding spices, many of which inhibit the bacteria and fungi that cause decay and mold. Because spicy foods also make you sweat, they may provide people in warm climates a way to keep cool.

Another reason why some people slather on the hot sauce is that they've lost some of their taste buds to the ravages of old age. In childhood, our mouths are populated with approximately

10,000 taste buds that renew themselves every week or two. But as we reach our fourth decade in life, the regeneration of taste buds begins to slow down. Individual taste buds don't weaken per se, but the reduction in their overall number explains why many people opt for bolder, spicier foods as they age; older people also tend to be on medications that can alter their sense of taste. Equally important, middle age is when we start to lose our sense of smell, which is a huge factor in our taste experience. (Incidentally, the loss of smell in the elderly is why they tend to apply way too much perfume.) Altogether, these dynamic changes in our aging senses tell us (1) it is a drag getting old, and (2) it is not unusual for tastes to change during one's lifetime.

Whether your affinity for the hot stuff is masochistic, cultural, or age-related, in the end biological limitations ultimately win the day. Foods with enough spice to make the devil gag are not to be toyed with. In 2014, Matt Gross gobbled down three Carolina Reapers in 21.85 seconds and suffered such severe heartburn that he thought he was having a heart attack for 12 hours. In 2016, a fellow who consumed a burger topped with ghost pepper puree nearly died. The puree caused such profuse vomiting that it ripped a hole in his esophagus, forcing him to eat through a feeding tube during his three-week recovery in the hospital. Vomiting is a common reaction to ingesting too much heat, because capsaicin induces increased mucous and intestinal contractions, the body's attempt to rid the offensive substance. Of course, if it isn't regurgitated, there is only one other way it gets flushed out, and it can burn almost as badly as when it entered. Seizures have also been reported in those who consume too many hot peppers in too short a time.

An analogous situation occurs with the refreshingly cool sensation you get when eating mints. Cold temperatures activate another thermal receptor on our cells called TRPM8, which tells our brain: Damn, this is cold! Menthol is a waxy chemical in mint plants like spearmint and peppermint that just so happens to be able to bind and activate TRPM8, too. Regardless of how TRPM8 is activated, our brain gets the same chilling message.

Thanks to science, we now know that certain foods and spices make us sweat or shiver by hijacking thermal receptors in our body. Although, let's face it: Science has yet to explain how anyone can stomach the Spice Girls.

Why You Can't Live Without Coffee

In the 1982 movie *Airplane II: The Sequel,* the flight attendant informs the passengers that their lunar shuttle has been thrown off course, asteroids are smashing into the hull of the ship, and the navigational system is down. But the passengers don't panic until she tells them one last bit of news: They're also out of coffee.

Many busy people throughout the world are rarely seen without a coffee cup in hand. Coffee is more than just a beverage to some people; it is like another limb. And just like a body part, our genes influence our preferences for coffee.

Coffee is a ubiquitous form of energy, an easy (and for most people, delicious) way to deliver the drug caffeine into our system. Caffeine stimulates us because it looks like another chemical we make called adenosine, which travels through our body as an indicator of our energy level. When we are awake, adenosine accumulates in our body, and eventually reaches a point when it binds enough of the adenosine receptors in our brain to say, "Yo, that's enough, time for sleep." Caffeine short-circuits this process by sitting in the chairs that are meant for adenosine; when caffeine blocks enough adenosine receptors, the brain does not get the message that it needs sleep. When enough caffeine tricks our neurons, our brain is fooled into thinking that there is an emergency and engages the "fight-or-flight" response by releasing hormones like adrenaline. Our concentration and memory get sharper, our heart beats faster, and our sugar reserves are released to boost energy levels.

Some people take pride in their coffee addiction, while others find they can't drink much of the stuff. These preferences may not necessarily be something you decided; rather, your DNA influences them. We've already covered one type of gene that can influence coffee intake: TAS2R38. Supertasters who can't stand the bitter flavors may not be able to tolerate stronger coffees, or might need to offset that bitterness with a lot of sugar and cream. But it takes a lot more than a TAS2R38 mutation to keep some people away from their caffeine fix.

Coffee preference goes beyond the taste buds, as caffeine affects people differently. A gene called CYP1A2 may explain why some people guzzle coffee like water with no ill effects, whereas others get jittery after a single cup. CYP1A2 encodes a so-called "cytochrome" enzyme in our liver that works to metabolize caffeine, among other things.

Not everyone's CYP1A2 cytochrome is created equal. Most people feel the effects of caffeine just 15 to 30 minutes after ingestion, and the drug's half-life is about six hours. (That's how long it takes for the body to eliminate half the amount of caffeine consumed. It's also why you might not want to have a lot of coffee with a six o'clock dinner: 50 percent of that caffeine will still be buzzing your system at midnight when you are trying to catch some z's.) However, people who have a certain variant designated CYP1A2*1F are classified as slow caffeine metabolizers. Their CYP1A2 enzyme is a bit of a slacker, and does not process the caffeine as quickly. In practice, this means that the drug stays active in the body for a longer period of time. Not only does this amplify the stimulating effects of caffeine, but it also increases the person's blood pressure. Some studies have even shown that slow caffeine metabolizers are at increased risk for a heart attack and hypertension when ingesting excess caffeine.

Ever notice how most smokers also drink a lot of coffee? That's because the nicotine in their cigarettes activates the CYP1A2 gene, which in turn causes the coffee's caffeine to be metabolized at a faster rate. So smokers tend to get a shorter boost from their caffeine hit, prompting them to reach for a second cup of joe sooner than nonsmokers.

Our ability to process caffeine, and undoubtedly other types of drugs and food, produces inequity in mental and athletic performance. One study from 2012 showed that slow caffeine metabolizers only improved their time in a stationary bike race by one minute after having coffee, but fast caffeine metabolizers improved their time by four minutes. Do we now need to ban coffee or other caffeinated beverages at the Olympics?

Another potential explanation behind our varied preferences for caffeinated beverages may involve the different types of bacteria present in our gut. The evidence that gut microbes can influence caffeine metabolism comes from a beetle called the coffee berry borer. This pesky critter is a menace to anyone who farms coffee for a living as it eats coffee beans for breakfast. And for lunch. And for dinner. The coffee berry borer is the only known creature that actually lives on nothing but coffee—consuming the equivalent of a 150-pound person downing 500 shots of espresso. How it survives the lethal doses of caffeine it ingests had long been a mystery.

In 2015, microbiologist Eoin Brodie of the Lawrence Berkeley National Laboratory led a team of researchers who discovered that several species of bacteria, such as *Pseudomonas fulva,* which reside in the gut of coffee berry borers, are wizards at breaking down caffeine. The *P. fulva* bacteria bring a caffeine-detoxifying gene into the beetle that enables it to live off coffee beans. Although there is presently no evidence that humans possess caffeine-busting bacteria like this, such species have been identified in coffee makers. If these bacteria are ingested and become a part of our microbiota, they can conceivably affect the rate of caffeine metabolism in our body.

As researchers uncover additional genes and possibly bacterial species that play a role in how we process caffeine, we will gain further insight into why some people feel a faster or greater boost from coffee. And how caffeine affects the body certainly influences a person's taste for caffeinated beverages.

Why Milk Doesn't Do Every Body Good

Got milk? Or got tummy ache? Many people around the world are not free to consume milk or other dairy products because their DNA deprives them of the enzyme called lactase. Lactase, which is encoded by a gene called LCT, is able to break down the lactose sugar molecules in milk. If your body doesn't metabolize the lactose, bacteria in your gut will—but at a cost. When gut bacteria feast on lactose, they produce a lot of gas that can lead to bloating and some rather embarrassing noises (which infallibly decide to make themselves heard during first dates). The lactose sugar also causes water from your intestinal cells to flow into the gut via osmosis, forcing your body to release it the only way it knows how. This is why people who do not produce enough lactase can suffer very uncomfortable cramping or diarrhea after ingesting dairy products. If you don't express the LCT gene and make lactase, it tragically means milk shakes, ice cream, and sometimes even pizza are off the menu. One of my friends is so lactose-intolerant that he can't even listen to a cheesy love song without passing gas.

Unlike the other genes discussed here, lactose intolerance is not usually due to a defective LCT gene. Virtually everyone is born with a functional LCT gene, as lactase is essential for digesting mother's milk. For most babies, the LCT gene shuts down shortly after their mom says last call. But some human ancestors in certain parts of the world where milk-producing animals were being domesticated (mostly Europe, the Middle East, and South Asia) procured a DNA mutation that allowed their LCT gene to stay on indefinitely. In other words, lactose intolerance is normal for adults; those who can digest lactose are the mutants. Scientists refer to these milk-swilling mutants as "mampires" because they suck up the mammary gland fluids of other animals. Ancestral mampires who could keep cranking out lactase after weaning had a strong survival advantage because they could get nutrients from additional sources, such as milk from cows, goats, or camels. It was such an important benefit that the mutation to keep the LCT gene "on" spread like wildfire in regions where it arose about 10,000 years ago.

Why You Think the $40 Bottle of Wine Tastes Better

I am convinced that most descriptions of wine are actually "Mad Libs" filled in by a computer algorithm. Try it! This (name of wine) tickles the nose with its (type of wood) aroma, reminiscent of a (animal)'s (body part) during a (season) day in (exotic location). It makes an (adjective) entrance on the tongue, like (a holiday), surprising you with a burst of crushed (fruit) that gently melds into (color) (type of spice).

Many of us just do not understand the complex flavors of wine, and some would argue that people who do are fooling themselves. Studies have been conducted in which professional wine tasters are served with flights of wine that, unbeknownst to them, have several identical samples. Only about 10 percent of these wine tasters were able to rate those identical samples with the same ranking. Studies have also shown that most nonprofessional tasters (and even some cork dorks) can't distinguish cheap from expensive wine in blind taste tests. What might help explain these variations in our ability to appreciate the complexities of wine? One possibility is the disproportionate amount of supertasters found among wine professionals. In other words, your broccoli-hating version of TAS2R38 could open up a whole new world of wine swishing and spitting fun for you.

However, as any wine enthusiast will tell you, there is much more to the grape than just the taste—for example, the bouquet. Smell is indeed an important component of taste, and variations in our genes can lead to variations in our olfactory experience. But a 2016 study by food scientist María Victoria Moreno-Arribas at the Spanish National Research Council suggested that our mouth bacteria can also affect the aroma of a wine.

Everyone has different types of bacteria pitching camp in their mouth—our so-called oral microbiota. A dynamic oral microbiota not only explains why some praise a wine while others insist it is rubbish; it can also explain why professional tasters can be inconsistent at times. Our oral microbiota are populated with bacteria and yeast from things we eat, drink, and inhale, and it can change on a dime. A rinse with mouthwash, for example, pretty much wipes the slate clean.

Researchers in this study focused on a little patch of nerves at the back of your throat called "the retronasal passage." Sounds like a new Harry Potter book, but it's actually where taste and aroma converge, providing the means by which you can "smell" wine in your throat. Researchers found that the type and number of bacteria caused different gas molecules to be released after they were mixed with grape aroma precursors, which may help to explain why your taste experience can be so different from someone else's. Perhaps if you engaged in a long, passionate kiss, your oral microbiota would be more similar and you'd come to agreement about the quality of the wine.

We are not all born to be sommeliers. The genes that happen to be within our DNA and oral microbiota play a large role in whether we are able to distinguish countless different types of wines and vintages, or if we're just content choosing between some red or some white.

Another fascinating line of experiments suggests that our brain can turn its back on our sense of taste. For his Ph.D. dissertation, winemaker Frédéric Brochet had a bunch of enology (wine science) students literally seeing red—when they were actually tasting a white wine. Brochet put tasteless red dye into the white wine, and every pro described the taste of the white wine as a red wine.

A similar thing happens during blind taste tests of Coca-Cola versus Pepsi. If someone gives you two cups of soda, and only one of them is labeled as "Coke," you are most likely going to claim that the labeled cup tastes better than the unlabeled cup—even if that unlabeled cup also contains Coke. Go ahead, try that out with your friends! And, if you happen to have a buddy with a brain scanner, you can try to repeat the experiment performed by neuroscientist Read Montague at the Baylor College of Medicine in Houston, which showed how much branding can influence us.

In the cola wars, the Force is strong with the Coke brand. Montague's study showed that when cola fans drank Pepsi in a blind taste test, their brain's pleasure center was much more active than when they drank Coke—clearly, they enjoyed Pepsi more than Coke. However, when the same cola fans were told which brand they were drinking, almost all claimed that the Coke tasted better. Intriguingly, when they knew which cola they were drinking, a different part of the brain was active—one associated with learning and memory. Montague believes the mind was conjuring the successful branding of Coke, influencing the subject's tastes more than the actual flavor of the cola itself. It's as if the subjects had been blinded by branding and lost the ability to think for themselves.

This does not necessarily mean that wine (or cola) tasting is rubbish. But it does indicate how easily we can be fooled. Our brain is a biased organ that is filled with preconceptions; it will sometimes even dismiss evidence that goes against what it thinks is true (see Chapter 8). Why is it so lazy? Our brain takes these mental shortcuts because it needs a lot of energy—up to 20 percent of our body's supply—and it uses cerebral shortcuts to conserve this energy. In the blind beverage-tasting studies, the brain trusts our sense of sight and ignores the contradictory data the taste buds send.

To illustrate just how influential our preconceptions can be, consider the famous fudge experiment from the 1980s. In this study, subjects were asked to choose between two identical samples of delicious fudge; however, one was shaped into a disk and the other shaped into a very realistic pile of dog poo. The subjects were told that each plate contained the same perfectly safe and edible fudge. But the brain didn't care. People overwhelmingly opted for the sample that didn't resemble a poop emoji.

In 2016, psychologist Lisa Feldman Barrett at Northeastern University discovered that our preconceptions about how livestock are raised can affect our taste perception of the meat. Even though all the meat sampled was from the same source, participants claimed it looked, smelled, and tasted less pleasant when they were told it came from a factory farm instead of a humane farm.

Our perceptions even get in the way of our tasting the most basic of beverages: water. Penn & Teller did a bit on their hit show *Bullshit!* that had a "water steward" ask diners at a gourmet restaurant to taste several brands of bottled water from all over the world. The customers in the experiment remarked how different each of the waters tasted, describing variations in hardness, crispness, freshness, and purity. All of them were adamant that each brand was far superior to tap water. Unbeknownst to them, each fancy bottle they sampled was in fact previously filled with water from the same garden hose on the restaurant's patio.

How You Can Eat Something So Disgusting

The most talked-about activities on reality game shows like *Survivor* or *Fear Factor* are often the food challenges. In these competitions, contestants scarf down the most unappetizing things imaginable, including sheep's eyes, Rocky Mountain oysters (testicles), tarantulas, baluts (duck embryos), cow brains, mangrove worms—and, in a fitting way to end this list, horse rectums. Why do we get grossed out by these foods, some of which are delicacies in certain cultures, and not grossed out by a Twinkie, which many people in the world would not consider to be a real food?

Across different regions (or maybe even across the street) you are likely to encounter someone who savors flavors that you find repugnant. Some cheeses, like Limburger, smell like the sock lint picked out of a soccer player's toenail, yet some folks in Wisconsin love it so much they built a sandwich around it. In Japan, people snack on natto, a fermented soybean dish that smells like the socks our soccer player left in the bottom of the hamper for a month. Durian is a spiky yellow-green fruit from Southeast Asia that smells so foul it has been banned from being carried on public transportation. Typically cited as the worst-smelling food ever is Sweden's *surströmming*, a fermented herring that is such an assault on the senses that the government recommends you open the can while outdoors.

There could be genetic explanations behind our love for unusually odorous foods. But in addition to taste receptors, there is evidence that we can acquire palates very early in life, even while in the womb. An unborn baby does several shots worth of amniotic fluid every day during its tenure in utero, and scientists have shown that the amniotic fluid can be flavored like soup based on what the mother eats.

Biopsychologist Julie Mennella at the Monell Chemical Senses Center conducted a study in 1995 that had participants sniffing amniotic fluid taken from pregnant mothers who ingested either garlic or sugar capsules. Although sniffing a stranger's bodily fluids doesn't sound like the first thing most people would opt to do on a Friday night, the study established that a mother's amniotic fluid does indeed carry flavors into the gestating child. It is also well established that food flavors can be transmitted to nursing children through breast milk.

But the proof is in the pudding so to speak, and so Mennella conducted another study to confirm whether flavor preferences could be transmitted from mother to child. The study involved three groups of pregnant women: one that drank carrot juice every day during the pregnancy, one that drank carrot juice only while breast-feeding, and one that avoided carrots during pregnancy and while nursing. After the babies were born, they showed a clear preference for carrot-flavored cereal if their mothers consumed carrot juice during pregnancy or while breast-feeding. Children who were never exposed to carrot juice as a fetus or while nursing were more likely to grimace when first tasting it. This study is consistent with the idea that a mother can impart her flavor biases to her unborn child.

As mentioned earlier, the environment in the womb is a predictor of what is out there on the other side. If a mother lives in a veritable carrot land, it would serve the fetus well to develop a taste for carrots. Similarly, if a baby tastes something it never experienced from its mother's fluids before, it should hesitate and possibly reject the foreign substance as potentially toxic. This is not to say these children are destined to hate carrots; rather, they may just be pausing to make sure the unfamiliar substance doesn't make them sick. The passing of food preferences from mother to child in utero is believed to partly explain why some cultures can stomach foods that others wouldn't dream of putting into their mouths.

Living With Your Tastes

What is more personal and self-defining than our likes and dislikes, especially when it comes to food and drink? It's unsettling to think that our dietary desires can be thwarted by a DNA bouncer at the refrigerator door, but this knowledge is power. (For example, if you are going to make your poor supertaster kid eat broccoli, at least serve it with some honey mustard chasers!)

On a more serious note, folks with the supertaster TAS2R38 gene variant consume an average of 200 fewer servings of veggies a year, which may lead them down an unhealthy road or put them at higher risk for colon cancer. If you are a supertaster, be sure you get enough veggies that you can tolerate, or find ways to make the bitter ones more palatable. Roasting vegetables allows the sugars to caramelize, bringing out sweeter flavors that mask bitterness. Another potential problem that could plague supertasters is high blood pressure. Supertasters have a tendency to use excessive salt on their foods to mask bitter flavors their taste buds can't tolerate. On the plus side, supertasters are less likely to be overweight, because they are also more sensitive to overly sweet and high-fat foods. For those who are lactose intolerant, an increasing number of dairy-free food options are becoming available. For those who can't savor the complexities of wine, just rejoice in how much money you'll save buying Trader Joe's "Two-Buck Chuck."

New studies showing how food can (perhaps irreversibly) program a baby's DNA in the womb provide one of the more compelling reasons why we need to eat more intelligently. As we'll see later on, fathers are not off the hook either, as their life choices can epigenetically program their genetic contribution to babies, too.

Understanding our differences in taste should help foster a more peaceful mealtime. Next time someone doesn't like your cooking or the wine you recommend, cut yourselves some slack and remember: There is a biological explanation behind our taste idiosyncrasies. So let it go and talk about a good book instead.

» CHAPTER THREE «

MEET YOUR APPETITE

I can't stop eating. I eat because I'm unhappy, and I'm unhappy because I eat. It's a vicious cycle.

—Fat Bastard, *Austin Powers: The Spy Who Shagged Me*

When people hear the word "outbreak," it usually conjures up images of Ebola, a zombie apocalypse, or that silly 1995 movie featuring Dustin Hoffman chasing a monkey. But in the United States and several other developed countries, another type of outbreak is taking shape—one that is reshaping waistlines. Our clothes have so many X's on them that some might think they've walked into an adult bookstore instead of their closet. Larger ambulances with heavy-duty hoists are being built to accommodate obese patients. Hit TV shows like *The Biggest Loser* have made an "enlightening" game out of weight loss. Entire channels like the Food Network have been created that cater to our insatiable appetite for food, and it is difficult to check our social media streams without being tempted by a picture of some new pizza-fried-chicken-burger monstrosity.

We try to counter our gluttony by eating someone else's words in the form of diet books. Unfortunately, the buffet of fad diets all have one thing in common: None of them seem to work for maintaining a healthy weight. Tellingly, used bookstores are overstuffed with carrot juice–stained diet books, which frustrated customers quickly sell after realizing the plan is useless. Even the winning contestants on *The Biggest Loser* have been known to gain the majority of their weight back after leaving the show.

According to the Centers for Disease Control and Prevention (CDC), the U.S. obesity rate is ballooning to near 40 percent. Another third of the population is overweight. One in five children are clinically obese. Most of us know the serious health risks that stem from living large, such as heart disease, stroke, type 2 diabetes, and cancer. And yet we continue to lose the battle of the bulge. This battle isn't cheap, costing the country more than $190 billion in obesity-related health problems every year. As the Western diet and lifestyle spreads around the globe, so do cases of obesity. A disturbing 2017 report from *The New England Journal of Medicine* shows that more than two billion children and adults around the world suffer from health problems related to being overweight. What is happening to us?

Our modern lifestyle explains much of the obesity epidemic: We eat a lot more and move a lot less. Not only do we eat like food is going out of style, but what we eat is wildly unhealthy. We all know this; we've been told about the food pyramid and the importance of exercise since dodgeballs slammed into our heads in elementary school. Yet we're watching the old adage of "you are what you eat" materialize before our eyes, with most of us looking more like a Cinnabon than a celery stalk.

Why is it hard to be rational when it comes to our appetite and food choices? Is this simply a problem with willpower, or could other factors be at play?

Why You Crave the High (Calorie) Life

As mentioned in Chapter 2, our lustful affair with sweets dips way back into our evolutionary past, when things like Mountain Dew and Krispy Kremes were not yet invented. High-calorie energy sources were hard to find on the African savanna where our species emerged. Early humans who felt a strong craving for energy-packed foods like sweet fruits, animal fat, or honey would have a clear advantage over those who didn't, because they'd have fuel reserves for hunting, fighting, shagging, and yelling at their kids to stop drawing on the cave walls. To persuade our ancestors to seek out these foods, evolution tapped into our brain's punishment and reward system.

Our DNA crafted a brain that could feel pleasure and pain to make its survival machine competitive. A survival machine that doesn't eat will soon experience hunger pangs. After eating, these pangs are replaced with a feeling of satisfaction. High-calorie foods like cheesecake take us beyond satisfaction though, leaving us with a pleasurable sensation that is near orgasmic. Our brain experiences a reward when something sweet hits our lips. Whether that something sweet is a Hershey's Kiss or French kiss, the result is the same—a surge of dopamine, the neurotransmitter that juices the reward center in the brain and demands a repeat performance.

The things we eat taste good or bad because it is our body's primitive way of tracking what might be useful to put in our stomach. Our DNA made high-calorie foods taste so good that we'd often risk life and limb to obtain them. But today, we only risk a limb when the vending machine fails to drop our Twix bar. We're surrounded by sweets and fats, and the only effort it takes to acquire them is getting off the sofa to answer the door for pizza delivery. We live in a real-life version of Candy Land, minus the physical activity: a surefire formula for weight gain.

According to the American Heart Association, recommendations for added sugars are five teaspoons a day (80 calories) for a daily energy expenditure of 1,800 calories for an average adult woman or nine teaspoons a day (144 calories) for a daily energy expenditure of 2,200 calories for an average adult man. I know some people who put five teaspoons of sugar in just one cup of their morning coffee (probably supertasters!). An average 12-ounce can of soda has about eight teaspoons of sugar. Many people reach their maximum allowance of sugar for the whole day with just one beverage.

We tend to underestimate how much sugar we're consuming because it's a "hidden" ingredient in many foods that we don't normally consider to be sugary treats. These include pasta and pizza sauce, barbecue sauce, salad dressings, juice drinks, and even some "healthy" cereals, yogurts, and granola bars. Foods that are obnoxiously high in sugar content do not exist in nature, and our body is not built to handle them.

When we have Frosted Flakes and a chocolate chip muffin for breakfast, we get a jolt of sugar that needs to be processed faster than our body knows how. The pancreas goes into overdrive to make enough insulin to manage the surge of sugar flooding our body. Insulin is a hormone that helps get the sugar molecules into the cells that need it; the rest is stored as fat. The overproduction of insulin helps to clear the sugar from our blood efficiently, but often leaves us with that unpleasant "crash" that we feel coming down from a sugar high. Consequently, we find ourselves needing another snack just to stabilize the blood sugar levels we threw so far out of whack.

A similar situation holds true for fats and salt. Our body needs fats and minerals like salt to function properly, so we don't want to eliminate these ingredients from our diet. But just as we see with sugar, people probably don't realize how much excess fat and salt they are consuming in popular food items. If you take in 2,000 calories a day, you should have between 44 and 78 grams of fat a day. A single fast food dish usually takes us into that range. One Cinnabon is almost 40 grams of fat. A bowl of mac and cheese can be upwards of 60 grams of fat. Surprisingly, many fast food salads toss in 30 to 60 grams of fat. Even some frozen mocha coffees can pack up to 50 grams of fat.

The U.S. Department of Agriculture (USDA) recommends that healthy adults consume less than 2,400 milligrams (about one teaspoon) of sodium daily. However, Americans typically take in 3,000 to 4,000 milligrams of sodium each day. The bulk of this excess salt comes from processed foods, packaged foods, and restaurant foods; a single frozen meal or entrée at a fast food joint can meet our salt quota for the day, or even more. Ranked as the number one saltiest food in America, P.F. Chang's Hot and Sour Soup Bowl packs nearly 8,000 milligrams of salt (more than three times the limit for the day).

Our constant munching on salty snacks like chips, pretzels, popcorn, and nuts, can easily take us close to the salt danger zone. Less obvious sources of high sodium include some soups, vegetable juices, sauces, and deli meats. One tablespoon of soy sauce can have more than 1,000 milligrams of salt.

Food companies and restaurants want repeat business. Adding lots of sugar, fat, and/or salt into their foods is effective, because they work the same way as drugs of abuse. We're like Al Pacino in *Scarface*, except our face is buried in a pile of powdered doughnuts instead of cocaine. Because high-calorie foods fire up the brain's reward system in the same way as opioids, junk foods are technically an addictive substance (which makes the word "junkie" a very prescient term). Multiple studies have shown that sugar is even more addictive than cocaine. That is why cravings for unhealthy snacks can be overwhelming. And because sugar is as pervasive as GEICO ads, getting people to kick the habit is like trying to set up a drug rehab in a crack house.

We crave processed foods with lots of sugar, fat, and salt because the supply and demand was reversed millions of years ago. Once scarce and precious, sugar, fat, and salt are now ubiquitous and cheap. Unprocessed foods usually cost more than processed now, and you still have to put in the time to prepare them. This likely contributes to why obesity, once a rich man's disease, is now skyrocketing among the poor and middle classes. The disparity is summarized well in this candid statement made by Monica Drane, daughter of the man who invented Lunchables, the hugely successful processed meal kit for kids: "I don't think my kids have ever eaten a Lunchable . . . They know they exist and that Grandpa Bob invented them. But we eat very healthfully."

Why You Eat Too Much

Most processed foods play right into the worst of our dietary cravings, creating an environment that makes it extraordinarily difficult to resist junk food. But our cravings are only a part of the whole story. What controls the volume of food we eat?

You know when your smartphone needs to be charged because of its battery indicator. In the body, a series of hormones function similarly. When our stomach is empty, our gastrointestinal cells release ghrelin, the "hunger hormone" that travels to the brain screaming, "Feed me!" As we begin recharging our body with food, ghrelin levels fall and we begin to experience the sensation of being full. The contentment that follows a meal comes from the "satiety hormone" called leptin, which our fat cells release. This elegant hormonal cycle controls, through biochemical signals, when we eat and how much. Some people overeat because of genetic mutations that have disrupted this hunger/satiety hormonal system. This would be like a smartphone that can no longer sense how much power is left in the device.

The first genes linked to appetite control were identified in mice, which might come as a surprise. When is the last time you saw a fat mouse wobbling across your floor? Probably never. Mice with extra weight would find it challenging to scurry away from predators. But in 1949, scientists were stunned to find an obese mouse lounging around the cage with its lean littermates. Purely by chance, this mouse was born with a mutation in its leptin gene; it was used to develop a strain of obese mice named *ob/ob*. Because these mice no longer make leptin, they never feel full. Remarkably, if researchers injected leptin into an *ob/ob* mouse, it stopped overeating. When you give an *ob/ob* mouse a cookie, he's bound to want many more. When you give an *ob/ob* mouse leptin, he's bound to feel full.

In 1998, researchers found that mutations in the leptin gene also existed in several patients with morbid obesity. Another bizarre parallel between the *ob/ob* mice and these patients with obesity is related to reproduction. *Ob/ob* mice are sterile (unless corrected by dietary restriction or leptin supplementation); leptin-deficient patients with obesity never go through puberty. Produced by fat cells, leptin is an indicator of body mass. Despite a body's massive size, if leptin isn't being made, the brain isn't getting the signal that the body has enough fat reserves for reproduction. This also explains why leptin mutations are so rare in humans and mice.

As seen in the *ob/ob* mice, leptin replacement therapy succeeds in restoring these people with morbid obesity to a near normal weight (it also induces puberty and restores fertility). This sounds wonderful, but can *anyone* lose weight by taking leptin supplements? Sorry, the answer is no. Leptin therapy only helps the few people in the world who are leptin deficient. In fact, most people with obesity actually make leptin just fine. The problem is that the brain has become resistant to leptin's effects, no longer properly responding to the hormone. The fat cells are trying to tell the brain that it can stop eating now, but the brain isn't getting the memo.

As you might surmise, whenever we talk about a hormone (or neurotransmitter), there has to be a receptor for that signaling molecule—like a ball to a catcher's mitt. LEPR (leptin receptor) is the gene that works as the catcher's mitt for leptin, and mutations in this gene can also disrupt the leptin signal of satiety. Another type of obese mouse, called *db/db*, was found to have a mutation in its leptin receptor, a condition that is also seen in some people. A small number of individuals with obesity suffer from leptin receptor deficiency—they never get the signal that they're full—which makes it virtually impossible for them to stop overeating. The satiety signal keeps on knocking, but it can't come in.

Leptin also promotes the production of another satiety hormone called α-melanocyte-stimulating hormone (α-MSH), which helps provide that sense of fullness after a meal. The receptor that catches and responds to this hormone is called MC4R, and scientists have identified DNA mutations near this gene in folks with early onset obesity.

Strangely enough, just as leptin mutations can be linked to puberty problems, there is another curious connection between food and sex involving MC4R. Chemists have developed a variety of compounds that look like α-MSH, in hopes that they will bind to MC4R and help people with obesity feel full. Turns out that some of these so-called MC4R agonists have "erectogenic properties," meaning that they give rise to an appetite for sex. If these turn out to be weight loss drugs one day, men who normally have to loosen their belts after meals will still be doing so, but for different reasons. (Eating less and added exercise—a win-win!) But joking aside, this example highlights the challenge we face: The genetics underlying our appetite is complex, and researchers have a lot on their plate to digest before they can determine whether drugs can help us safely balance our caloric checkbook.

In addition to genes that regulate the hormonal aspects of feeding behavior, there are genes that regulate the elation we feel after eating a good meal. Researchers have found that people with a genetic variation called Taq1A are more likely to become obese, because it reduces the amount of dopamine receptors (the receptors involved in sensing reward) in the brain. People possessing this variant need to overeat to feel the same dopamine reward most others experience with lesser amounts of food.

It cannot be overstated how important studies like this are. Genes can regulate our feelings. For some people, appetite is not about self-control—it is beyond their control. Is it right to judge or admonish people with ravenous appetites because of the way their DNA was cooked up?

Even if DNA built you a brain that is able to properly regulate food intake, other things can go wrong that are beyond our control, leading to changes in behavior that result in weight gain. Scientists have documented numerous examples of patients gaining or losing weight due to a brain tumor or concussion. Additionally, some people who have brain implants that deliver electrical pulses to control involuntary movement disorders (such as Parkinson's disease) suddenly start binge eating. Similarly, when certain brain regions in mice are stimulated, the mice start rapidly gorging on their food within seconds. These studies underscore the important role of the brain in our eating habits.

Ongoing studies to identify the specific brain regions controlling feeding behavior may lead to novel treatments for weight problems. In 2018, neurobiologist Charles Zuker at Columbia University was able to manipulate specific neurons in the brains of mice such that their natural desire for sweets, and their aversion to bitter flavors, were both erased. Imagine one day having your brain rewired in a way that makes broccoli more appealing than chocolate cake!

How Your Parents Influenced Your Appetite

There's no question that mishaps in certain genes can disrupt appetite control and metabolism in some folks. But genetic variation cannot explain the rapid surge in obesity in recent decades. It has been estimated that mutations in single genes like those discussed previously account for less than 10 percent of the obesity epidemic. Rather than DNA mutation, perhaps an epigenetic mechanism is at play—something in our environment may be altering gene expression in ways that facilitate obesity. Supporting this idea, there is evidence that the food we eat can alter gene expression in ways that influence appetite, metabolism, and susceptibility to disease.

One of the most striking observations reflecting the importance of diet on gene expression comes from biologist Randy Jirtle's work at Duke University Medical Center. Through the study of a gene called agouti in mice, he and others have shown how critical a mother's diet is during pregnancy. What does this agouti gene do, you ask? It puts blond highlights into a mouse's hair. Wait, what? How we go from blond highlights to a mothers's diet to changes in the child's appetite requires a little more explaining.

From the instant sperm and egg collide, a newly formed team of genes rolls up their sleeves and get to work on building a baby. An elaborate chain reaction begins to unfold, where different waves of genes are turned on and off in a highly ordered progression. From day one, the environment provided by a mother's womb begins to program the genome of the baby for survival on its own. Nature assumes that the environment a mother is experiencing while pregnant is the same environment that baby will have to navigate. Therefore, the level of activity for some of a baby's genes is programmed *before* birth in an effort to prepare the child for life outside the womb. This is called fetal or prenatal programming.

Picture a mouse in your mind. You probably pictured an energetic, small and furry rodent with a brown coat. Most mice do indeed look like this; however, sometimes a mouse will be born with a yellow coat and grow to be a large furball—looking more like a Tribble from *Star Trek*. These yellow-haired mice are predisposed to diseases including obesity, diabetes, and cancer. The difference between the normal and yellow mouse is that the yellow one never switched off its agouti gene. If you examine a normal mouse's hair up close, you might see that the middle part is yellow, which denotes when its agouti gene was on. The brown parts of the hair, flanking the yellow, denote when the agouti gene was off. In some mice, the agouti gene is never active and they have hair that is completely brown. In yellow mice, the agouti gene is never turned off, producing its golden coat.

You'll remember our previous discussion of the satiety hormone α-melanocyte-stimulating hormone (α-MSH). Like many hormones, α-MSH has multiple effects on the body; in addition to providing that feeling of fullness after eating, α-MSH darkens hair. The agouti protein blocks the binding of α-MSH from its receptor on hair follicle cells, so the hair stays blond. But these blonds do not have more fun. Unfortunately for the yellow-haired mice, that agouti protein also interferes with the binding of α-MSH to the MCR4 receptor on brain cells, causing the mice to overeat because they are no longer getting the signal that they are full.

But, you may be wondering, the agouti gene is present in all of these mice, so why are some brown and lean and others yellow and fat? What gives rise to the variation in coat color and future health prospects is how much of the agouti gene is active and when.

As you would expect, yellow, obese mice typically give birth to pups that grow up to be yellow and unhealthy just like their mother. But here's the jaw-dropping discovery: If you tweak the diet of a yellow mother during pregnancy, she gives birth to brown, lean, healthy mice! What kind of sorcery is this? The trick was to supplement the mother's diet with nutrients that enhance DNA methylation (such as folic acid, betaine, vitamin B_{12}, and choline), the chemical change that turns genes off. One of the genes that is silenced through dietary-driven DNA methylation is agouti. With that agouti gene silenced, the pups born to yellow obese mothers had brown fur and did not overeat because the agouti protein that could block the α-MSH receptors was never made.

This remarkable study has several implications related to appetite. First, the DNA of a developing fetus is being actively programmed before birth to gear up for life in the mother's environment. This fetal programming is based, in part, on signals from the mother's diet as a proxy to what her surroundings are like. Second, what the mother eats while pregnant can have a lasting impact on her offspring by altering the methylation of the baby's DNA. Dietary supplements beyond what health professionals recommend should be treated with caution, because we have no idea what most of them can do to fetal DNA. Third, although you inherit the genes of your mother and father, it does not mean you will express them like your mother or father. Yellow agouti mothers may be fat, but her pups are not destined to share this fate. Of course, this concept could work in reverse, too. Further study of genes and their epigenetic control during development may guide a pregnant mother to program her unborn child for optimal health.

Before fathers-to-be get too comfortable behind their pork tenderloin and triple-decker brownie sundae tower, they should be aware that their eating habits can also affect their future children through epigenetic programing of their sperm. A 2010 study by pharmacologist Margaret Morris at the University of New South Wales in Australia found that male rats given a high-fat diet had female pups that were plagued with insulin problems, which coincided with weight gain and increased fat. A closer look at the daughters of these junk food–eating fathers revealed abnormal expression of over 600 genes in the islet cells of their pancreas, the cells that are in charge of making insulin.

What about humans? To find out if weight might affect gene expression in human sperm, biologist Romain Barrès at the University of Copenhagen in Denmark examined DNA methylation patterns in sperm samples donated by men who are both lean and obese. In a 2015 study, his team found that more than 9,000 genes were methylated differently between sperm from men who were lean versus obese. That's a lot of genes, and among them were some of the usual suspects already linked to appetite control, like MC4R. These alterations in gene expression could lead to permanent changes in how the brain is wired for appetite control, potentially explaining why some people find it nearly impossible to change their eating behaviors. In the same study, Barrès and his colleagues also looked at DNA methylation in sperm before and after the men with obesity underwent gastric bypass (bariatric) surgery to lose weight. DNA methylation changes were detected in genes like MC4R one year after the bariatric surgery, suggesting that it is possible to reverse these modifications to sperm DNA.

In contrast to the studies done in rodents, the extent to which these DNA methylation changes in the sperm of men with obesity affect their children has not been studied. However, all of these studies provide a plausible mechanism by which your father's eating habits could have influenced your own, way back when half of your DNA was still swimming around in his nether regions. Sorry, dads-to-be, but it seems that the mother is not the only one who needs to live a healthy lifestyle prior to conceiving.

Why Sugar Might Make Your Life Sweeter, but Shorter

What you are fed as an infant and toddler may have long-lasting effects on your eating behavior as an adult. To help study how certain activities affect an entire life span, scientists often use model organisms like fruit flies and tiny roundworms called *C. elegans*. Because worms and flies go through their entire life in about 15 or 90 days, respectively, scientists can study how various things affect the development of these organisms from birth to death. To study an entire life span in humans would take the entire life span of the scientist, but it is really hard to get tenure if you don't publish until you're in your 70s.

We can use short-lived creatures to ask questions like "What happens to young adults who eat lots of junk food? Do they turn out OK?" The news is not so good. A 2017 study performed by biologist Nazif Alic at University College London revealed that the life span of young flies fed a high-sugar diet for three weeks was cut short by 7 percent. Unfortunately, their life was abbreviated even after the high-sucrose diet was switched to a healthy one. That's right . . . flies that were fed lots of sugar during their youth died prematurely even if they started adulting and ate right.

When Alic's team dug into the mechanism behind sugar's ability to irreversibly shorten life spans, they found that sucrose interfered with a transcription factor called FOXO. Humans have these same factors. Transcription factors regulate the expression of multiple genes, so interfering with one disrupts an entire network of genes, like falling dominoes. The FOXO gene network makes many proteins that keep our cells in good working order, so it makes sense that organisms would live longer with this network engaged. Too much sugar dampens the activity of the FOXO gene network, and alarmingly, these changes appear to be unshakeable even after his sugar-fed flies start a healthy diet.

These studies suggest that excess sugar accelerates aging, making life literally short and sweet. What a sugar buzzkill! It is important to keep in mind that there are still unresolved questions about how this all works, and studies in humans are very limited at present. Although many studies show that excess sugar damages our health, it may be premature to conclude that permanent damage is done to children who grew up on a Willy Wonka diet. Nevertheless, it is advisable to go ahead and make sensible dietary changes for yourself and your children sooner rather than later.

How Your Gut Bacteria Might Be Swaying Your Appetite

A growing number of scientists believe that our intestinal microbiota could also be influencing our appetite and tipping our scales. It is more than just a "gut feeling," as studies with germ-free mice have shown. Germ-free mice are born and raised in a sterile environment with no microbes living in or on them. No other creature on Earth is this pure, not even June Cleaver. Although these sterile mice may be living the dream through the eyes of a germophobe, this level of purity has negative consequences. Germ-free mice are scrawny, have defects in their immune system, and respond inappropriately to stress. The abnormalities seen in germ-free mice demonstrated a surprising idea: Intestinal bacteria are not merely vagabonds passing through, but they play important roles in our health and well-being. Life without them is not life as we know it.

Scientists wondered what would happen if they colonized the intestines of germ-free mice with microbes. To do this, researchers had to play a "dirty trick" of sorts on the mice. They harvested some material from the cecum (the beginning of the large intestine, or colon) of a normal mouse and spotted it onto the coat of the germ-free mouse. When the germ-free mouse went to groom itself, it unwittingly lapped up the bacteria-laden cecum goo. After grooming, the germ-free mouse became a germ-ridden mouse.

The skinny germ-free mice that inoculated themselves with bacteria from the gut of a normal mouse gained considerable weight within two weeks. In an amazing display, this bacterial transplant made the scrawny mice look as normal as the ones that "donated" the bacteria. The weight gain was not due to increased appetite, because the formerly germ-free mice actually ate *less* food after getting bacteria from a normal mouse. What are these bacteria doing that helped the germ-free mice get to a normal weight? They brought new genes on board that are good at digesting the complex plant carbohydrates present in mouse chow. The mice could now extract more energy from less food, thanks to the improved digestion service the bacteria provided.

Researchers then got a wild idea to test if all gut bacteria were created equally. What would happen to the germ-free mice if they were fed bacteria that came from an obese donor, like the *ob/ob* mouse? The germ-free mice fed bacteria from an obese *ob/ob* mouse gained *lots* of weight, and soon resembled their obese donor. Germ-free mice gained nearly 30 percent more fat when fed bacteria from normal mice, but upward of

50 percent more fat when fed bacteria from obese mice. This result suggests that intestinal bacteria are indeed different in lean versus obese mice, and that they have effects beyond the entrails that impact weight gain.

Inspired by these pioneering studies, a group of researchers have become the "Lewis and Clark" of intestines, and began to map the different species of bacteria inhabiting mice of different sizes. The National Institutes of Health has initiated a similar effort to document the microbial denizens of humans called the Human Microbiome Project. The findings are still coming in, but just as explorers don't find the same animals in the forest as they do in the desert, scientists are seeing differences in the types of bacteria inhabiting lean versus obese guts. For example, obese mice have more bacteria called Firmicutes and less bacteria called Bacteroidetes. Both Firmicutes and Bacteroidetes house many different types of bacterial species that are commonly present in the gut of mammals. Firmicutes bacteria are able to extract more calories from food compared to Bacteroidetes, which makes it easier for a body to pack away fat reserves.

Most studies support a decreased Bacteroidetes:Firmicutes ratio in human obesity as well. An unbalanced diet is associated with an unbalanced microbiota. A 2009 study by the pioneering microbiota scientist Jeffrey Gordon at Washington University compared the intestinal bacteria of obese and lean twins and found differences in 383 bacterial genes. In participants with obesity, 75 percent of these genes were from bacteria called Actinobacteria and 25 percent were Firmicutes. No genes from Bacteroidetes were detected in the participants with obesity. In contrast, 42 percent of these genes in lean individuals were from Bacteroidetes. Many of these bacterial genes found to differ between individuals who are obese versus lean are involved in metabolism.

Additional evidence supports an intimate connection between our diet, weight, and the composition of our microbiota. Biologist Paolo Lionetti at the University of Florence compared the microbiota in people living in a European city to those in a rural African village. As you might suspect, children in Florence, Italy, tend to be heavier than the lean children in the African village of Boulpon in Burkina Faso. Interestingly, Lionetti found that kids in Florence who largely ate a "Western diet" rich in carbs, fat, and salt housed mostly Firmicutes bacteria. Children in Boulpon are believed to still be dining on foods that our early human ancestors likely ate: mostly fruits and vegetables, with protein coming from occasional meat, eggs, or insects (termites in this case). The microbiota of Boulpon's kids had far more Bacteroidetes and large portions of two species not found in the Italian kids at all: *Prevotella* and *Xylanibacter,* both experts in digesting the fiber found in plant-based foods.

The types of bacteria housed in our gut change quickly in response to our diet. Lean people who start to eat a higher-calorie diet see their Firmicutes bacteria rapidly increase at the expense of their Bacteroidetes bacteria, which causes them to extract even more calories out of their food. Eating junk food could create a vicious cycle that quickly shifts the balance of gut bacteria toward the types that promote fat formation.

Like the belly of a mouse with too many Firmicutes, the plot thickens. It was pretty cool to watch bacteria from a fat mouse help a scrawny germ-free mouse start living large. But what happens if we feed it bacteria from humans? In 2013, Gordon's team harvested intestinal bacteria from twin humans, in which one was substantially heavier than the other. Amazingly, when they transplanted gut bacteria from the twin with obesity into germ-free mice, those mice became fat. Gut bacteria from the twin that was thin did not induce significant weight gain when put into the germ-free mice.

In this same study, Gordon and his colleagues set up a twist whereby the germ-free mice fed bacteria from the thin twin were placed into the same cage as those fed bacteria from the heavy twin. To understand what happened, you need to be aware of a little habit mice have that most of us find a bit unsavory: They eat one another's poo. But because they do, they basically perform their own fecal transplants when caged together. Here is the astonishing result: The mice colonized with bacteria from the heavy twin were able to stay thin if they ingested poo excreted by mice harboring bacteria from the thin twin. It seems the bacteria from the thin twin has some sort of superpower that prevents weight gain.

Now before you text your bony friends for a little "favor" with a poo emoji, note that when this experiment was repeated with the mice being served high-fat chow instead of their usual plant-based chow, those bacteria from the lean twin lost their superpowers and could no longer prevent weight gain in the mice seeded with bacteria from the obese twin. In other words, getting lean bacteria is not enough. Healthy eating habits are required. These results tell us that gut bacteria and diet influence one another in ways that likely affect metabolism, but many details still need to be sorted out before we can manipulate the system.

How You Might Be Able to Change Your Cravings

How different types of bacteria gain a foothold in your gut relates to your appetite. It seems that bacteria down below send chemical signals to your brain that influence your cravings for the kinds of food they need to propagate and outcompete other types of bacteria that want to move onto their turf. There is a circle of microbial life in your belly: What you eat affects your microbes, and what they eat may affect what you eat.

Just like any other living creature, the bacteria swarming around in our intestines are competing for space and nutrients. Lucky for us, millions of years of evolution have forged symbiotic relationships between our DNA and theirs, achieving a "scratch my back and I'll scratch yours" kind of existence. We provide them with a place to live and a free lunch, and they will make some vitamins for us and help keep nasty pathogenic bacteria and fungi at bay. But all bacteria want to optimize their growth, so they could have evolved ways to trick our brain into sending them the foods that help them flourish.

The types of bacteria in someone's gut reveals insights into what they eat. Compared to most people in Japan, farmers from the midwestern United States are not as well equipped to digest seaweed, a major constituent of a sushi-based diet. People of Japanese descent have acquired unique Bacteroides bacteria capable of digesting and extracting nutrients from seaweed. Others can enjoy sushi, of course, but they might not digest the seaweed as efficiently as Japanese diners. Studies like this suggest that the composition of our microbiota are not only influenced by what we eat today, but also by what our ancestors ate.

Rather than on seaweed, some species of bacteria thrive on sugar, still others on fat; these bacteria could be manipulating you to crave junk food. Perhaps you can break the spell, however, by rebelling against these junkies in your gut and eating healthy, unprocessed foods. If so, you'll soon have more bacteria in your gut making you hungry for a cranberry walnut salad instead of a bacon cheeseburger and fries.

What you eat (or fail to eat) influences your bacterial demographics. Processed foods tend to have a paucity of fiber, which is extraordinarily unusual in the dietary history of our species. Fiber has long been a mainstay in the human diet, but most of us get next to none of the 25 to 35 grams we need each day. This lack of fiber not only makes a bowel movement feel like you're passing bricks, but it also puts you at elevated risk for colon cancer. The lack of fiber starves certain bacteria that bestow major health benefits upon us.

Studies on mice that were given a fiber supplement saw a boost in their *Bifidobacteria,* yet another type of bacteria in our guts. *Bifidobacteria* are the original "probiotic" and have long been suspected of being beneficial to intestinal health. Probiotic means "for life" and refers to microbes that exert a positive effect on health. Such microbes are plentiful in fermented foods and dairy products, which have been recommended for thousands of years to treat gastrointestinal problems. The idea was popularized in the late 19th century after scientist Élie Metchnikoff noted that Bulgarians, who seemed to live long and hardy lives, frequently consumed sour milk or yogurt.

The levels of a bacterial species called *Akkermansia muciniphila* also soar in response to the fiber. *Akkermansia* bacteria are much more prevalent in people who are lean, and are virtually absent from people with obesity, type 2 diabetes, or inflammatory bowel disease. *Akkermansia* promotes turnover of the mucus layer that lines the gut walls, which acts as an important barrier to prevent leakage. When stuff leaks from the gut, it can cause inflammation in the body, including inflammation within fatty tissue, thus leading to weight gain. High-fat diets decimate *Akkermansia,* but nourishing your microbes with fiber might reward you with a better ratio of fat to body mass, decreased inflammation, and reduced insulin resistance.

And that's not all fiber can do for you! One study in mice showed that dietary fiber was able to reduce inflammation in the lungs caused by allergies. The study showed that in addition to shifting the microbiota away from Firmicutes to Bacteroidetes, fiber was processed into short-chain fatty acids (SCFAs) that act on other parts of the body. In this case, the SCFAs produced from the digested fiber went to the lungs and altered the immune response in ways that reduced allergic inflammation. Conversely, mice fed a low-fiber diet showed increased allergic airway disease. Recall Lionetti's previously mentioned study, which compared the microbiota of urban Italian and rural African children. He also found that the bacteria in the African children correlated with higher SCFA levels compared to the Italian children. He speculates that this may help explain why Africans suffer less inflammatory disease.

Knowing that bacteria break down fiber and other foodstuffs into chemicals that can affect multiple systems in the body, scientists have proposed two potential tactics that microbes can use to manipulate our appetite. One, bacteria can produce chemicals that go to our brain and initiate cravings for foods that the bacteria need to grow. Two, bacteria can produce chemicals that make us feel like crap until we eat foods the bacteria need. So not only are our bacterial overlords driving our appetite but they may also be causing our mood swings. More on this in later chapters.

So the next time you sit down at a table, remember that you are not eating for one; you are eating for trillions of little critters teeming throughout your gut. You may find yourself eating when you're not even hungry because these bacteria are always asking to be fed. And in the modern world where food is aplenty, it is all too easy to give the bacteria what they want, and it is not always what you need.

Why You Don't Have an Appetite for Exercise

Despite exercise being the best medicine for so much of what ails us, our excuses to avoid sweating could fill a book larger than the collective works of Sue Grafton. Although our excuses range from absurd (I already carried a heavy box of doughnuts up the stairs at work today) to legitimate (a dog bit off my big toe during my run yesterday), there is little debate that we only get a tiny fraction of the physical activity our body needs.

It's fairly obvious why. Exercise hurts. It's uncomfortable. It makes us stink. Why spend precious time making ourselves miserable when we can kick back with a chocolate martini and watch another episode of *Cupcake Wars?* Truth is, exercise is not as much of a chore as we make it out to be; the problem is that we've pampered ourselves to such extremes that pressing a remote control is now considered too hard because we can just issue voice commands to stream the next episode of *Diners, Drive-Ins and Dives*.

In our modern world, light exercise seems like a Herculean effort because we've made everything else so effortless. Science has thoroughly documented the plentiful benefits of exercise: It builds strength, increases our energy level, lowers blood pressure, reduces stress, helps with depression, and staves off weight gain and the myriad health problems associated with being overweight. Exercise has even been shown to extend life, enhance memory and learning, and slow mental decline. So why aren't we all getting off our butts?

Our genes may provide part of the answer. Studies of twins have revealed a genetic component that influences our tendency to work out or stay put. Some people are born with genes that build bodies that are less amenable to certain types of exercise. Kathryn North at the University of Sydney in Australia conducted a study showing a significant link between a gene called ACTN3 and athletic performance. This protein, which is found in fast-twitch muscle fibers that generate force at high velocity, is enriched in sprinters and strength exercisers. Others have a mutation that stops this protein from being made, and these people tend to be endurance athletes. So whether you prefer to run short or long distances, or lift weights instead of doing aerobics, may boil down to the type of muscles your genes build.

There's no question that professional athletes work and train very hard, but many also have a genetic edge to thank for their gift. Sometimes the genetic benefit is obvious, as in the case of basketball players so tall they have to watch out for low-flying aircraft. But sometimes the genetic benefit is more cryptic. For example, Olympic hero Eero Mäntyranta made cross-country skiing look easy. We know today that high-endurance events *were* easier for him, thanks to a mutation he carried in his erythropoietin receptor (EPOR) gene. This mutation allowed him to produce a larger number of red blood cells than normal, granting him a superhero-like ability to deliver oxygen to his starving muscles way faster than his competitors could. Which brings up an interesting question: Do you think it's fair if you had to compete against someone who had this condition? Or if you did, should you be allowed to take erythropoietin (EPO)—the hormone cyclist Lance Armstrong admitted to taking—to raise your red blood cell count?

Another reason some people enjoy exercise more than others could be because they experience a greater reward in the brain following physical exertion. Variations in genes involved in the dopamine reward pathways in the brain may be linked with someone's level of physical activity. When people feel rewarded by exercise, they are more inclined to go for it. If you don't feel an innate reward after hitting the gym, you could find other ways to treat yo'self for completing a workout.

Any fan of "Garfield" knows that we're not the only animals that like to slack off. Frank Booth at the University of Missouri noticed that some of his lab rats exercised much more often on the cage wheel than other rats, which avoided the wheel like Homer Simpson would avoid a bike. His team selectively bred these animals to make exercise-loving rats and exercise-hating rats, and then compared gene expression in their brains. Some of the gene expression variations support a difference in the dopamine reward pathways between these rats, bolstering the idea that some individuals genuinely experience a reward following exercise and others do not.

The motivation to exercise can be dramatically curtailed by eating junk food, creating a double whammy against good health. Studies have shown that the Western diet strongly correlates with laziness and depression, leading researchers to conclude that people with obesity are not necessarily heavy because they are lazy or lack discipline, but because the junk food changed their mood and behavior. Studies performed in rats support this idea. Rats fed an unhealthy diet do not just get fat; they are also significantly less motivated to perform reward tasks.

Scientists are also investigating the differences in the microbiota between athletes and nonathletes. Many of the same themes are emerging that we've seen in healthy versus non-healthy eaters. A 2017 study led by computational biologist Orla O'Sullivan at the Teagasc Food Research Centre in County Cork, Ireland, compared the gut microbiota of male rugby players to that of men who are sedentary. The microbiota in athletes were not only more diverse, but were also replete with health-promoting *Akkermansia* bacteria. In addition, despite the wear and tear on their muscles, the athletes had lower levels of inflammation, which could be attributed in part to their healthier microbiota.

The microbial species found in people who routinely exercise produce butyrate, which has strong anti-inflammatory properties. In another 2017 study, Maria del Mar Larrosa Pérez at European University, Madrid, found that the microbiota of women who engaged in modest exercise (three to five hours a week) was markedly different than inactive women. Even light exercise was associated with an increased presence of beneficial bacterial species like *Akkermansia*.

Food for Thought

In 1966, Brian Wilson of the Beach Boys sang, "I Just Wasn't Made for These Times." In truth, none of us are made for these times. We're engineered to live as hunter-gatherers who eat real (unprocessed) food and get plenty of physical activity. But we've created an environment so contrary to that of our ancestors that we are now facing an epidemic. For most people who are struggling with a muffin top or the onset of dad bod, keeping fit is a simple matter of education in making better food choices and getting off the couch. But for others, weight control is a serious lifelong challenge caused by a complex interplay of genes, environment, and possibly microbiota.

Consider the Pima Indians, who are now spread out between Arizona and Mexico. Existing as hunter-gatherers in or near the desert for thousands of years, the Pima evolved "thrifty" genes that made them masters of extracting calories from their scant food sources. Today, a striking difference is evident for anyone who visits the Pima in Mexico, who maintain a labor-intensive agrarian lifestyle, and the Pima in Arizona, who adapted the Western lifestyle of eating lots of processed food and moving very little. Take a guess at the difference you'd notice. The Arizona Pima now have one of the biggest problems with obesity in the world, with 60 percent suffering from type 2 diabetes. The Pima of Mexico have no such problems. Because the Pima in both areas have minimal genetic variation, the environment is a clear culprit contributing to the plight of the Arizona sect. An abundance of crap food combined with diminished physical activity has turned their once thrifty genes into a life-threatening liability.

The Pima illustrate how critical it is to eat and move like our ancestors did. They also illustrate the importance of genes in the weight loss equation, as some people are simply born to be more metabolically thrifty. Hundreds, perhaps thousands, of genes affect our appetite, metabolism, and our capacity to exercise. We must acknowledge that genes in our DNA and our microbiome play a major role in eating habits and weight gain, and accept that some things about our body shape are beyond our control. Science has shown that people with the same diet and lifestyle gain wildly different amounts of weight because of our genetic makeup. Writing for the science journal *Nature,* Stephen O'Rahilly beautifully summarizes the current state of our knowledge: "The growing evidence that humans can be genetically hard-wired to become severely obese should eventually lead to a more widespread realization that morbid obesity is a disease requiring further scientific research, rather than a failure of will-power requiring sanctimonious moral opprobrium." Fat-shaming is not only a cruelly repugnant approach, but many studies have proven it ineffective and damaging to people's health and well-being.

As we have seen, numerous biological factors—factors not always within our control—make it extraordinarily difficult for some to control their appetite. If willpower has worked for you, congratulations. But know that self-discipline is also influenced by genetics. Science will eventually solve our obesity problem. But until then, compassionate support and encouragement is a much better way to help ourselves and others meet realistic health goals.

» CHAPTER FOUR «

MEET YOUR ADDICTIONS

There's really no plausible
medical reason why I
should still be alive.
Maybe my DNA could say why.
—Ozzy Osbourne

Nowadays, you can sequence your genome in a weekend for about a thousand bucks. But would you believe that the first human genome sequenced took 13 years (1990 to 2003) and cost $2.7 billion?

Back in those days, when Harry Potter had just started his onscreen adventures at Hogwarts, getting your genome sequenced was a rare privilege. Among the first people to remove the invisibility cloak from their DNA was James Watson, one of the scientists who helped solve the structure of DNA in 1953, and Craig Venter, who was instrumental in making the Human Genome Project happen. Steve Jobs was also one of the first to have his genome sequenced (which I imagine lab techs referred to as the iGenome). What other luminaries did scientists reach out to for secrets their DNA held? Stephen Hawking? Marilyn vos Savant? Former President Barack Obama? That guy who won 74 games in a row on *Jeopardy?*

Nope. Scientists wanted Ozzy Osbourne.

Born in 1948, John Michael Osbourne answers to several names, including "Ozzy," "Prince of Darkness," and "Godfather of Heavy Metal." Ozzy rose to stardom with Black Sabbath in the 1970s, and then went on to a wild, and wildly successful, solo career. But Ozzy's music has often been eclipsed by his legendary drug and alcohol binges. So why would researchers want to peek inside Ozzy's genes?

Truth be told, Ozzy is a remarkable human specimen. He's constantly struggled with addiction to multiple vices (cocaine, booze, sex, pills, burritos), tirelessly toured and partied for half a century, withstood a near daily blast of "30 billion decibels" from stage speakers, and survived reality television (although he was taking up to 25 Vicodin pills a day at the time). His immune system was so weakened from drugs and alcohol that he once falsely tested positive for HIV.

Living one week of the Ozzy lifestyle would easily kill most of us, so scientists couldn't wait to get their latex-gloved hands on this Iron Man's DNA sequence. What death-defying genes could Ozzy possibly have that would allow someone to survive cocaine for breakfast and four bottles of cognac a day for decades?

In 2010, scientists at Knome, Inc., read the DNA diary of a madman and discovered that Ozzy is indeed a genetic mutant. Among some of the more intriguing things spotted in his DNA was a never-before-seen mutation near his ADH4 gene, which may explain why he could guzzle a liquor store in a day. ADH4 makes a protein called alcohol dehydrogenase-4, which breaks down alcohol. A mutation near ADH4 is likely to affect how much of the protein is made. If Ozzy's body is built to detoxify alcohol much faster than normal, it might help explain why his liver hasn't exploded.

Ozzy also possesses variations in genes linked to addiction, alcoholism, and the absorption of marijuana, opiates, and methamphetamines. All told, his DNA diary revealed that he is six times more likely than the average person to have alcohol dependency or alcohol cravings, 1.31 times more likely to have a cocaine addiction, and 2.6 times more likely to have hallucinations caused by marijuana.

Ozzy, who claimed that "the only Gene I know anything about is the one in KISS," was fascinated by the results. And although the variants found in his genome are tantalizing, the truth is we don't know enough about these genes yet to build a comprehensive picture that shows us why this man has an addictive personality or why he is still reasonably healthy after abusing his body for more than 50 years. Frankly, the data are just a "Blizzard of Ozz" at the moment. Addiction is a complex behavior, but research is revealing that our genes, along with other biological factors outside our jurisdiction, can for some conspire to make life a living hell.

Is Alcoholism in Your Genes?

Alcoholism includes the following four symptoms: craving, loss of control, physical dependence, and tolerance. The National Council on Alcoholism and Drug Dependence estimates that one in every 12 adults suffers from alcohol abuse or dependence in the United States alone. Americans spend nearly $200 million *a day* on booze, and about 100,000 people die each year from alcohol-related causes, such as drunk driving, suicide, falling down stairs, or thinking they can fly.

Alcohol addiction is clearly a serious problem, but I'm not trying to cast alcohol as the devil's nectar. The important question is why some people can't stop imbibing when they know they should. The vast majority of people enjoy alcoholic beverages because they like the taste, want to unwind a bit, or have to visit their in-laws. Most people use alcohol, but some abuse it. Why?

For the longest time, the stereotype has been that people with alcoholism are weak individuals who simply lack the willpower to stop at one drink. By the same token, some people who can drink like the chaps in *Mad Men* with minimal consequences chalk it up to their ability to put mind over matter. Neither of these notions is true. Science has demonstrated that your ability to control alcohol intake and how much it affects you have a significant genetic component. This is why people in the know treat addiction as a disease. The craving for alcohol can be just as potent as that for food or water; the craving can be so intense that it trumps virtually everything else in life, including family, friends, and even one's own welfare. People with addiction can have as much trouble turning down a drink as a starving person turning down a meal.

The National Institute on Alcohol Abuse and Alcoholism states that genes are responsible for about half of someone's propensity to develop alcohol addiction. But as with Ozzy's genome, there is rarely a single gene that fully explains this complex behavior. Indeed, numerous genes have been linked to alcohol dependence. The first one we'll look at relates to why people like to hit the pub after a stressful day at work.

A 2004 study by geneticist Tatiana Foroud at the Indiana University School of Medicine linked a gene called GABRB3 to alcoholism. This gene makes a subunit of the brain cell receptor that recognizes gamma-aminobutyric acid (GABA), a so-called "inhibitory" neurotransmitter that tells the brain to calm down. GABRB3 has also been linked to other conditions involving disruptions in normal brain activity, such as epilepsy, autism, and savant syndrome.

The discovery that GABRB3 is associated with alcohol abuse gives credence to the theory that the disease stems from an overactive brain. Due to its sedative properties, alcohol relaxes hyperactive neurons, serving to dam the raging rivers in the mind. Because alcohol addiction usually begins before the brain is finished developing in the early 20s, people with overactive brains learn to associate alcohol with relief. Learned behaviors acquired at this age can be extraordinarily difficult to change, because they are essentially embroidered into the brain's tapestry.

A 2015 study conducted in mice connected a gene called NF1 to alcohol dependence. Because NF1 influences the production of GABA, this study bolsters the idea that the GABA signaling pathway plays an important role in alcoholism. When researchers mutated NF1 in mice, the mice with the mutant gene drank more alcohol than normal mice. (Mice enjoy alcohol and drugs, too, which makes them useful model organisms for studying addiction behavior.)

To make sure the connection between NF1 and alcoholism isn't just a mouse thing, the team examined the NF1 gene in

9,000 people. Consistent with the mouse studies, they found an association between NF1 and the onset and severity of alcoholism. But just like studies with the GABRB3 receptor, more research is required to figure out precisely how these genetic changes result in alcoholism. One clue may be that mice with mutant NF1 do not produce as much GABA. Without GABA, the brain has trouble settling down, which may prompt someone to drink more for the calming effect.

Scientists are devising clever ways to identify additional genes associated with alcoholism. A 2014 study led by hepatologist Quentin Anstee at Newcastle University exposed mice to a chemical mutagen that alters their DNA (this is loosely analogous to David Bruce Banner zapping himself with gamma radiation in an attempt to mutate his weakling DNA into Hulk DNA). The researchers then selected which pups with the mutant gene had a preference for alcohol-laced water instead of normal water. The scientists traced the DNA defect back to the GABA receptor once again, but a different subunit of the receptor this time (called $GABA_A R$ β1). The mutation showed that a minor tweak to a single subunit of just one type of brain receptor heightened electrical activity, putting their minds into a constant state of overdrive. The electrical frenzy in the brains of these mice needs to be quelled; water won't do the trick, but alcohol will. These mice not only preferred the alcohol-laced water, but also continued to drink it even after they were three sheets to the wind.

Studies consistently show that alcohol addiction problems are not failings of character; rather, alcoholism has a genetic basis. The evidence shows a biological explanation for this unusual and often self-destructive behavior. Figuring out the genetic issues that give rise to substance abuse will offer much more effective rational ways to combat this disease than blaming and shaming the victim.

Why Some People Just Say No

Genes that govern how the body deals with alcohol or other drugs also influence whether someone is more prone to become a substance abuser. For example, some people, particularly those of East Asian descent, experience rapid flushing and a quickening heart rate when consuming alcohol. This is commonly referred to as Asian flush or Asian glow, but the more inclusive name is alcohol flush reaction, or AFR. People with AFR possess a genetic variant that impairs production of an enzyme that helps metabolize (break down) alcohol.

In the liver, alcohol is broken down to acetaldehyde (still toxic) and then to acetate (nontoxic). In those with AFR, alcohol is converted to acetaldehyde just fine, but then the acetaldehyde is not broken down efficiently. The buildup of toxic acetaldehyde causes blood vessels to dilate, which produces the redness and heat we call flushing. Excessive acetaldehyde can also cause headaches and nausea. The uncomfortable sensations associated with drinking prompt some to lay off the sauce, making people with AFR less likely to suffer from alcoholism.

Why did AFR arise in some people during the course of evolution? According to one study, the genetic mutation that causes AFR originated about 10,000 years ago in southern China, when people there began farming rice. In addition to rice as a food source, it could also be fermented to make alcohol, which was likely used as a disinfectant or preservative. Some of the more curious in the bunch probably ingested some to see what would happen, discovering alcohol to be a blessing and a curse. Researchers have speculated that the alcohol intolerance among these ancient people may have conferred a survival advantage by dissuading carriers from drinking excess quantities. The same principle underlies the use of disulfiram as a treatment for alcohol abuse. Disulfiram causes drinkers to experience the same unpleasant reactions of AFR when they consume alcohol, discouraging them from hitting the bottle.

Drugs have different effects on different people, largely based on what users have in their genetic toolbox to process the substance in question. The way our bodies respond to various drugs also explains why marijuana is far less dangerous than crack. A 2014 study by neuroscientist Pier Vincenzo Piazza at Université Bordeaux found that rodents produce a hormone called pregnenolone in response to marijuana. You're probably wondering how scientists roll tiny joints for rodents to smoke, but they actually inject them with tetrahydrocannabinol (THC), the psychoactive chemical in weed that gets people stoned by binding to cannabinoid receptors in the brain. Made from cholesterol, pregnenolone blocks THC from binding those receptors, reducing the impact of the drug. If this response is conserved in people, it may explain why marijuana does not appear to have a lethal dose and does not easily lead to addiction.

In another example, about 20 percent of Americans have a mutation in a gene called fatty acid amide hydrolase (FAAH), which makes an enzyme that breaks down anandamide, the so-called "bliss molecule" that your body produces naturally to decrease anxiety by binding your cannabinoid receptors. People with the mutated FAAH have more anandamide in their brain all the time; they not only tend to be calmer and happier than others, but they are also less likely to use marijuana because it simply doesn't do much for them.

Finally, some people have mutations in an opioid receptor in their body that shields them from addiction to opiates like morphine and OxyContin. These discoveries demonstrate that genetic differences explain why some people don't feel the buzz, and why they are less likely to become addicted.

Why Alcohol Hits Some Harder Than Others

We all know that one friend who falls off the bar stool in a giggling frenzy after just one drink. Alcohol may be hitting our lightweight friends quickly because they are literally light weight, of petite size. Or, they may be in possession of a variant of a gene called CYP2E1, which has been associated with getting people drunk faster on less alcohol.

The CYP2E1 gene produces another type of enzyme that is important for degrading alcohol to acetaldehyde. For the 10 to 20 percent of people who possess a particular variant of CYP2E1, those first few drinks leave them feeling more inebriated than the rest of the human population. Considering that people who respond strongly to alcohol are less likely to fall into alcoholism (like those with AFR), the CYP2E1 variant could be another type of survival advantage that helps carriers stay within their limits. We should be more understanding of our lightweight and flushing friends. They're not wimps; their bodies are generating more toxins at a faster pace.

In addition to our lightweight drinking buddies, there's always someone in the crowd more likely to act like an idiot after pounding a few. (You know, the people who start singing Leo Sayer's "You Make Me Feel Like Dancing" from the top of a stranger's car.) Such friends might be genetic mutants. Psychiatrist Roope Tikkanen at the University of Helsinki found that people who suffer from wildly poor judgment after a couple drinks have a mutation that results in a reduction of serotonin receptors, specifically one called serotonin 2B receptor. Serotonin is a neurotransmitter that regulates mood and behavior; therefore, it would follow that the loss of a serotonin receptor would short-circuit the ability to control behavior (in other words, the neurons in your brain are not getting the memo). Therefore, carriers of the mutation leading to less serotonin 2B receptor are more likely to engage in impulsive and aggressive behaviors while under the influence, and are also more prone to mood swings and depressive symptoms while sober.

Why It's So Hard for Some People to Stop Drinking

Like other drugs, alcohol causes the release of dopamine in the brain. Recall that dopamine is the neurotransmitter associated with reward behaviors; it makes you feel good and motivates you to repeat said behavior. When drugs induce dopamine, it prompts users to jump on that euphoric biochemical train again. And again. And again.

The power of dopamine to zombify us was illustrated in a *Star Trek: The Next Generation* episode called "The Game," in which the crew was seduced into submission by an alien species after becoming addicted to a video game the aliens introduced. The game became addictive because it directly stimulated the pleasure center of the brain fueled by dopamine, literally rewarding someone for playing. This is not unlike the popular video games on our planet, like Candy Crush or Angry Birds. Studies have shown that video games stimulate dopamine release, paving the way for addictive game play. In recent years, on rare occasion, young men have died binge playing video games for 24 or more consecutive hours. Virtually any activity that a person finds rewarding will release dopamine and has the capacity to become addictive, so it's best to choose your activities wisely.

Alcohol and other drugs are foreign chemicals that are processed by the body. If the body keeps seeing alcohol over and over, it reacts by making the liver work overtime to increase the number of enzymes to get rid of it. The body's attempt to resume normalcy is why drinkers build up tolerance to alcohol, which means they must ingest more and more to get the same feeling of satisfaction. To novice drinkers, one shot may produce a buzz. But after a few weeks of drinking, it will take two or three shots to reach that buzz, because their liver is processing the alcohol more efficiently.

After prolonged drinking, people who drink heavily will need to consume alcohol just to feel normal. To compensate for its sedative effects, our brain chemistry adapts to make more neurotransmitters that activate neurons to excite them again. If alcohol intake suddenly stops, the brain is no longer being sedated, but those excitatory neurotransmitters are still cranked up to 11. This is why people undergoing withdrawal experience the shakes, anxiety, and restlessness.

Because the brain takes time to recalibrate to the lack of alcohol, many people suffering withdrawal symptoms resume drinking just to calm down. The excess alcohol that needs to be consumed starts to wreak havoc on other bodily systems including the liver, kidneys, and stomach. Benzodiazepines like Xanax and Valium are sometimes administered to people undergoing alcohol withdrawal as a means to replace the effects of alcohol with a medicine that increases the anxiety-reducing neurotransmitter GABA. The administration of benzodiazepines can be better controlled than alcohol intake, and often helps restore the proper balance between excitatory and inhibitory activities in the neurons.

Alcohol interacts with numerous other systems in the brain, and genetic variations can exist in any of them, explaining why responses to alcohol and the tendency to become addicted vary so widely. Traditionally, scientists have uncovered genes associated with increased drinking, but a 2016 study led by Gunter Schumann at King's College London revealed a gene that may explain why some people know their limits. A variant in the gene that makes a protein called beta-Klotho was found in approximately 40 percent of study participants who show a decreased desire to drink alcohol.

The beta-Klotho protein is a receptor in the brain that catches a hormone called FGF21, which the liver secretes when it is processing alcohol. The scientists believe that beta-Klotho may be involved in cross-talk between the liver and the brain, a type of SOS that there is too much alcohol in the liver. When the team genetically engineered mice without beta-Klotho, these mice drank more alcohol. Such a feedback mechanism is analogous to how the satiety hormone leptin tells the brain when the stomach is full (see Chapter 3).

Studies like these suggest that people's ability to know their limits with alcohol is not necessarily due to stronger character or superior self-discipline. Rather, they were lucky to be born with a more effective liver-brain communication system. Moreover, understanding the biology of addiction will lead to new therapies to combat alcoholism. For example, when Schumann's team administered the FGF21 hormone to mice, it suppressed their preference for alcohol.

By imaging the brains of people with addiction, scientists have established that prolonged use of alcohol and other drugs produces significant, enduring changes. Brain scans of people with long-term drug abuse reveal alterations in regions important for impulsivity control, judgment, and decision making. Once these types of changes set in, it becomes increasingly difficult to break the cycle of addiction.

As many experts and people with addictions argue, being enslaved by a drug is not a lifestyle normal people would sign up for. It is not that people with addictions don't want to change, but their brain has been damaged. Like a pancreas that can no longer make insulin, an addicted brain can no longer generate the chemicals that regulate self-control. We don't persecute diabetics for their hormonal shortcomings, so is it fair to persecute people with drug addiction for theirs?

Epigenetic mechanisms are currently under investigation to reprogram the brains of people with addiction. Drugs can produce stable epigenetic changes to DNA, which can come back to haunt users even after years of sobriety. Recall that histone proteins interact with DNA and can be chemically modified (for example, though acetylation) to affect gene expression (see Chapter 1). In a 2017 study, neuroscientist Yasmin Hurd at the Center for Addictive Disorders at Mount Sinai found that the longer a person used heroin, the more histone acetylation activates genes like GRIA1, which has been implicated in drug-seeking behavior. Importantly, Hurd's team was able to decrease drug use in rats addicted to heroin by giving them JQ1, a compound that interferes with histone acetylation. These studies suggest that it may be possible to repair the epigenetic damage in brains of people with addiction and prevent relapse in the future.

Surprisingly, even our gut bacteria may play a role in addiction and the ability to resist relapse. From a 2014 study, microbiologist Fredrik Bäckhed at the University of Gothenburg reported that differences in how well people with alcoholism recovered in rehab could be linked to the composition of their intestinal microbiota. Almost half of the people with alcoholism in the study had an altered microbiota linked to a condition called "leaky gut" syndrome, which refers to seepage of intestinal biochemicals and debris out of the gut and into the body. It's as nasty as it sounds because out in the body where they don't belong, these biochemicals cause inflammation in organs and tissues. Bäckhed's study found that people with alcoholism and leaky guts craved alcohol more than those with alcoholism and normal intestinal bacteria. It is hoped that microbial supplements could one day be given to people recovering from addictions to prevent leaky gut, which in turn might help them recover by reducing cravings.

Why Don't You Do Drugs?

If drugs make us feel so good, perhaps the question really isn't "Why do you do drugs?" but "Why *don't* you do drugs?" Despite the myth, the vast majority of people who try drugs do not get hooked. A study done in 2008 showed that only 3.2 percent of people who try alcohol become addicted. Ever been warned that a single dose of crack, meth, or heroin is enough to convert anyone into a lifelong junkie? Not true for most people. Although drug experimentation is common, drug addiction is rare, with only 10 to 20 percent of users becoming dependent. This is by no means an invitation to try risky substances; the misery can be catastrophic if you fall into the minority group that gets hooked instantly. The point is that understanding why most people don't become enchained by a drug might help us break the chains binding those who do.

As we've seen, some people have a genetic predisposition that helps prevent them from falling into addiction. Some people react badly to drugs, feel little effect, or have genes that help them gauge their limit. But in addition to our genes, the environment plays a major role in whether a person stays clean and sober after trying drugs.

A predisposition for addiction may be programmed into children before they are even born, based on their mother's environment. The stress hormone cortisol, made by the adrenal glands, tends to accumulate in people living in nerve-racking conditions. Exposure to high levels of cortisol while in the womb can offset a baby's stress control system, which can emerge as a risk factor for addiction later in life. This idea has been validated in rats; rat pups that were subjected to prenatal stress in the womb are more prone to drug-seeking behavior and addiction when they grow up.

Unsurprisingly, children born to mothers who do drugs while pregnant are at greater risk of becoming addicted themselves. Drugs taken by a pregnant mother may adversely affect an unborn child's DNA through epigenetics. In a 2011 study, Yasmin Hurd showed that pregnant rats high on marijuana give birth to pups that have been epigenetically programmed to express fewer dopamine receptors. With their dopamine response dampened, these rats grew to have heightened risk-taking and pleasure-seeking behaviors, making them more vulnerable to addiction. In the same study, Hurd's group validated that response in humans: Prenatal marijuana exposure also diminished dopamine receptor expression in the brain. It seems biologically unfair, but children do pay the price for the mother's behavior during pregnancy in the form of fetal programming.

Early childhood trauma, referred to as an adverse childhood experience (ACE), is another major risk factor in developing addiction. A study conducted in Sweden that monitored participants from 1995 to 2011 showed that a single ACE, including parental death or assault, before the age of 15 is enough to double the chances of addiction. The more ACEs, the greater the odds of substance abuse later in life.

It is no secret that drugs are highly prevalent among people who are poor and disenfranchised. People with nothing to lose who need an escape from their grueling surroundings are more likely to experiment with drugs to take a break from all their worries. While at Simon Fraser University, psychologist Bruce Alexander famously demonstrated in 1977 the extent to which a healthy environment can thwart substance abuse and addiction. Like many other researchers, Alexander used rats to study drug abuse; rodents react to cocaine, alcohol, and other dopamine-triggering substances as much as we do, and their attraction to these substances can escalate into full-blown addiction. Lab rats trained to press a lever to get cocaine can become so addicted that they won't even take a break to eat. That alone should give you a new appreciation for how abnormally intense cravings can be for those with addictions.

It struck Alexander that all drug studies at the time were being performed on rats kept in small cages all alone with nothing to do. He wondered what would happen if the rats were put into a more stimulating environment, which led him to build "Rat Park." Sparing no expense, he created a veritable paradise for the rats, giving them plenty of space to roam and objects to explore. He put males and females in the park together and even set up areas where they could snuggle and start a family.

Alexander then spent six weeks getting a group of rats hooked on morphine before releasing them into either Rat Park or a dreary isolated cage. Both environments contained morphine-laced water and regular water. Remarkably, the overwhelming majority of rats living it up in Rat Park switched to the plain water. In contrast, the poor caged rats stuck with the morphine-laced water. On average, our behavior is not much different than the rats in this experiment. If people are lucky enough to be in an environment that naturally stimulates their dopamine reward response, then most won't seek out unnatural ways to do so.

More recent work by neuroscientist Carl Hart at Columbia University also suggests that people reject drugs when presented with better options. In 2013 he recruited people addicted to crack to stay in the hospital for several weeks. Every morning, they were given some crack. Later in the day, they were given a choice: more crack or five dollars. Hart found that many of the people opted for the cash. When he upped the amount to $20, every person picked cash over another hit. Hart believes that other "competing reinforcers" besides cash, such as sports, music, the arts, or clubs, help steer predisposed people away from the road to addiction by plugging their need for stimulation into a healthier outlet.

Similar findings have arisen for an addiction of another kind: food. Instead of wasting money on fad diets, one 2017 study demonstrated that sustained weight loss is better achieved when overweight people are paid to reach their health goals. Monetary rewards appear to be a much more effective way to incentivize people to address their addictive behaviors. Studies have shown that paying people a little bit of money to stop smoking now saves society huge amounts of money for their care later on.

Having competing opportunities is especially critical during adolescence, the period when most people are introduced to addictive behaviors that will come to dominate their lives. If you survive the teen years without developing an addiction, the odds are in your favor that you won't develop one later. Similarly, many people with addictions resolve them on their own, often without intervention, by their early 30s. What is it about the adolescent years that makes us particularly susceptible to addiction?

Many people don't realize that during puberty, the brain undergoes dramatic developmental changes that are not complete until the early 20s. Even though it goes against modern societal norms, the teenage years are when we are primed to make new life. Although it may not be advocated in those Paleo lifestyle books, the truth is that our Paleo ancestors had their first child far earlier than we do now. Therefore, in an evolutionary context, puberty gives us license to reproduce our DNA; as a result, we tend to engage in more exploratory risk-taking behavior to seek out and impress a mate. (By this rationale, a young man who ventured out to wrestle a bear for the tribe was more likely to make the young girls swoon—assuming he did not become the bear's lunch.)

As anyone with a teenager knows, areas of the brain that regulate self-control and good judgment have not fully developed yet. The "teen brain" reacts more strongly to receiving rewards, which can be a double-edged sword, because this makes learning easier, but can also lead to riskier behavior. Most of today's teens don't wrestle bears. But they will wrestle with opportunities to try drugs and alcohol, motivated by the same evolutionary drives. Especially vulnerable are teens that don't live in a human equivalent of Rat Park.

In *Unbroken Brain,* Maia Szalavitz argues that addiction is a learning disorder acquired during that critical window of adolescent brain development. Szalavitz proposes that some teens try alcohol or other drugs as a coping mechanism, which will cement itself as a learned behavior in a minor subset of kids predisposed to addiction. In other words, the brain has been rewired to associate the drug with relief and the ability to feel normal. Despite being a maladaptive means to deal with stress, these behaviors learned during adolescence are particularly difficult to unlearn, making flipping the script more difficult.

Although environment is a critical driver of addiction, it's never the whole story. The previously cited studies seem to suggest that if we create stimulating environments for everyone, then all of our drug problems would evaporate. And although there's no question that engineering smarter and more humane environments would help many people stay sober, it's naive to think that they will be the cure-all for everyone. There are always exceptions to general rules. Many impoverished people avoid the perils of drugs, and many substance abusers live in splendor. As we've learned, people exhibit biological features that influence the likelihood of addiction to a certain substance.

Now we'll examine personality traits that make resisting temptation harder for some.

Why You Engage in Risky Business

Some people seem to be born with a death wish: an addiction to an adrenaline rush, often a prelude to alcohol and drugs. These people engage in risky behaviors like skydiving, surfing the Pipeline, betting on the Cleveland Browns, or drinking raw milk. It's not your imagination when you see recklessness running in a family; studies of twins support a genetic component for thrill seeking. People experience different degrees of pleasure in response to the same stimuli, due to genetic variations in their dopamine system. Those who don't feel a normal level of pleasure may be inclined to take greater risks to fill this void. I'm happy as a clam reading a book about the ocean's creatures, but others wouldn't be happy unless they were wrestling sharks.

One of the high-profile genes linked to risk-taking behavior is DRD4, which encodes a type of dopamine receptor associated with our motivation to obtain and derive pleasure from a reward. Variants in the DRD4 gene have been associated with increased novelty seeking and risk taking, and may have been selected for long ago when our ancestors began to migrate out of their native land of Africa. Although some of our restless ancestors craved adventure and a change of scenery, others said, "No, I'm good," opting to stay home and chill. When the DRD4 gene was examined in native peoples around the globe, the people farthest away from the birthplace of our species were most likely to harbor the risk-taking variant.

Being an exploratory sensation seeker can have its advantages, but the trait has some downsides as well. DRD4 variants are found in people diagnosed with attention deficit/hyperactivity disorder (ADHD); incidentally, one in four people who have an addiction problem also have ADHD. DRD4 variants have been associated with alcoholism, opioid dependence, unsafe sex, and gambling. It should be noted, however, that research correlating DRD4 variants with novelty seeking have not always been in agreement and most certainly depend on interplay with other genetic or environmental factors. For those who might have a genetic predisposition for daredevilry, what steers them toward Mount Everest versus a mountain of cocaine? That's where environmental factors may make all the difference. Kinesiologist Cynthia Thomson at the University of British Columbia, who found a DRD4 variant enriched in daring skiers and snowboarders, has stated, "Given no healthy outlets for their sensation seeking, such individuals might turn to more problematic behaviors, like gambling or drugs."

Do you make hasty decisions or know someone who does? Some people are more impulsive than others. They don't deliberate too much about the options before them, nor do they take the time to weigh the potential consequences. It's the kind of behavior that makes the lives of the boys in the *Sons of Anarchy* so exhilarating and so miserable. Over a dozen genes have been linked to deficits in impulse control, and most involve different aspects of our nervous system, including neurotransmitters like serotonin, dopamine, and norepinephrine, among others. Importantly, every single one of the genes associated with impulsivity is also associated with alcoholism or some other addiction. These findings bolster the case that addiction has more to do with our DNA fiber than our moral fiber.

Whether or not we are born with a predisposition toward impulsive behavior can have dramatic consequences for our entire lives. The "marshmallow test" conducted by psychologist Walter Mischel at Stanford University famously illustrated this. In the 1960s Mischel told a group of preschool children that they could eat one marshmallow now or they could eat two marshmallows if they waited 15 minutes. A third of the children gobbled up their single marshmallow right away. Another third held out for about three minutes before giving in to temptation. The final third was able to stick it out the whole 15 minutes to get two marshmallows. These results validate that self-control (or lack thereof) is a behavior that can be observed very early in life. However, the really interesting part came 30 years later, when Mischel conducted a follow-up study on the children who took his marshmallow test decades before. Turns out that on average, the impulsive kids who couldn't wait to eat that marshmallow grew up to have alarming problems that generally weren't seen in the kids who could delay gratification. Preschoolers who devoured the marshmallow straightaway tended to earn lower SAT scores and had lower-paying jobs; additionally, they were more likely to be obese, jailed for criminal behavior, or addicted to drugs.

It should be mentioned that Mischel's marshmallow study has been roasted for having too small a sample size and being fraught with confounding variables. Newer studies have suggested that poverty and other social disadvantages encourage children to jump at short-term rewards, whereas children from households that have more education and money find it easier to delay gratification and wait for the larger reward.

Despite the caveats, you may be tempted to test your children if they can resist temptation in this manner. But don't fret if they fail it. This knowledge is power, and you can focus on teaching your children strategies to delay gratification and master self-discipline. (Other studies have shown that just being told what happens to people who fail the marshmallow test helps people practice better self-control. So, you're welcome.) You can also teach your kids the simple strategy that the children who were able to resist temptation used: distracting themselves from thinking about the reward.

A 2016 study led by neuroscientist Huda Akil at the University of Michigan demonstrated an epigenetic component to risk-taking behavior that is linked to addiction as well. Her team bred rats to be "high responders (HR)" or "low responders (LR)." HR rats are more inclined to be novelty seeking and impulsive, whereas LR rats don't take chances. As you might expect by now, HR rats are much more prone to becoming addicted to cocaine than the LR rats. One epigenetic difference between the HR and LR rats results in decreased expression of a dopamine receptor. The lack of this dopamine receptor in HR rats is believed to dampen the pleasure response, causing them to be thrill seekers. By way of comparison, it doesn't take much stimulation for LR rats to feel pleasure, so they don't have to work so hard to get their kicks. Supporting this idea, after the HR rats become addicted to cocaine, their dopamine receptor levels rose to match the levels seen in nonaddicted LR rats. These findings not only establish that epigenetic differences contribute to risk-taking behavior, but also confirm that drugs alter epigenetic marks on DNA.

Startling evidence produced by gastroenterologist Premysl Bercik at McMaster University in Hamilton, Canada, also shows that our microbiota can influence whether we are high rollers or like to play it safe. One of the most striking findings concerning this point came when a timid strain of mice suddenly became adventurous explorers after receiving gut microbes from a brave strain of mice. Could the types of bacteria in our gut that day really influence our willingness to take risks in life? (Instead of a lousy medal, perhaps the Wizard of Oz should have given the Cowardly Lion a fecal transplant using the stool from a normal lion.)

Meanwhile, another microbe—one that is far more stealthy and sinister—may be manipulating our behavior in a number of ways. An astonishing third of us have a single-celled parasite called *Toxoplasma gondii* lurking in our brain. The parasite only causes symptoms in immune-compromised individuals, but will persist for the rest of any infected person's life in the form of latent tissue cysts, which like to camp out in the brain, among other places. Presently, there is no cure for this latent stage of toxoplasmosis.

Studies by parasitologist Jaroslav Flegr at Charles University in Prague have shown that people infected with this parasite (which we can get from cats or from consuming contaminated food or water) display distinct personality profiles that may increase impulsivity and risk-taking behavior. Consistent with this observation, a 2015 meta-analysis (a study of studies) found a positive correlation between *Toxoplasma* infection and addiction. Increased risktaking may, in part, also explain studies showing that people infected with *Toxoplasma* are nearly three times more likely to be in a traffic accident. Do you know someone who likes to get freaky in the bedroom? Studies show that *Toxoplasma* may also be the reason for that, too. (Imagine, a cat parasite turning people into sex kittens!)

How this parasite may manipulate the brain is a question under intense investigation, and it's likely that the parasite affects different people in different ways. As *Toxoplasma* invades the brain and sets up parasite cysts inside neurons, it could do structural damage or change brain chemistry. *Toxoplasma* secretes a large array of proteins into the host organism's cells, most of which have yet to be characterized. Additionally, *Toxoplasma* infection alters the host immune response, which can also influence behavior.

What a Long, Strange Trip It's Been

You may never have thought about it this way, but just about everyone is or has been addicted at one point to caffeine. Sure, caffeine is mild compared to hardcore drugs, but the fundamental principles are the same. We enjoy the jolt of energy that caffeine brings, but soon we can't seem to function without it. We get tired and cranky. Many people are complete ogres until they have had their morning coffee. After a little while, we find ourselves having a second or third cup because one just doesn't cut it anymore. Try to stop and you will be tortured with fatigue, headaches, and irritability. It's easier just to brew another pot and keep the habit going. If asked to give up the coffeepot, many would say you'd have to pry it from their cold, dead hands.

It's the same basic cycle those with other addictions, but involving substances that are much more difficult to quit. Perhaps we can use this common ground to reshape our approach for helping people with addiction issues. The addiction is punishing enough, and further punitive action has proven to be an abysmal failure that has needlessly ruined the lives of many good people. The true crime of people with addiction is having the wrong genes in the wrong place at the wrong time. With better education, we can prevent more people from doing drugs in the first place. With a better understanding of the biology behind addiction, we can develop effective treatments. With a better idea of the genes predisposing people to an addictive personality, we can screen for people who may be at risk. With a better understanding of the environmental forces that drive addiction, our resources can be spent more intelligently.

In the misguided war on drugs, the *victims* have been put in the crosshairs. We assume people predisposed to addiction should have enough willpower to just say no, but science has shown this assumption to be horribly wrong. Most people do not, in fact, become addicted to alcohol and drugs. But a minority of people have genes that put them at risk of escalating experimentation to addiction. Along the way, the brain changes and makes it virtually impossible to break the cycle without professional intervention. We need a war on addiction, not a war on drugs, and certainly not a war on those who are addicted.

In *The Anatomy of Addiction,* psychiatrist Akikur Mohammad points out the flaws in our current methods to treat people with addiction. Despite their familiarity, the famous 12-step addiction programs are profoundly unsuccessful, helping only 5 to 8 percent of participants stay sober. Many of these programs do not have a licensed addiction therapist or medical professional; many also ban the use of medicines like the combination of buprenorphine and naloxone, which helps stop cravings for opiates, or other medications that ease the pain of withdrawal. Some even forbid the use of medication that a person with an addiction may need for a mood disorder. (Nearly half of individuals with severe mental disorder are affected by substance abuse.)

To win the war on addiction, Mohammad has proposed an evidenced-based strategy that includes biomedical, psychological, and sociocultural components. In addition to taking medications designed to fight addiction, people with an addiction should receive behavioral self-control training and aversion therapy, as well as community reinforcement. Consistent with the studies of neuroscientist Carl Hart, contingency management, which involves granting rewards for sobriety, is emerging as one of the most effective tools against addiction. Nicotine is an ideal example of a harmful drug whose use has greatly diminished in recent years, thanks to educational campaigns, health insurance incentives, and the development of medicines to help smokers quit.

It is important we understand that addiction is an incurable, chronic brain disease that requires lifelong management. Blaming, shaming, and punishing victims simply does not work. And yet society can't seem to break its own addiction to this outmoded and ineffective approach.

» CHAPTER FIVE «

MEET YOUR MOODS

If you were happy every day of your life, you wouldn't be a human. You'd be a game show host.

—Veronica Sawyer, *Heathers*

"Mork calling Orson . . . come in Orson!"

The moment I heard those words as an eight-year-old kid, I was hooked. Like many others, I fell in love with the comic genius of Robin Williams. In his television debut with *Mork & Mindy,* Williams played an alien sent to Earth to study human behavior (much as we're doing in these pages). The show launched a legendary career that ended abruptly in 2014, when Williams committed suicide. The tragedy shocked the world: How could such a delightfully funny man—one who could light up the room with his smile—be so sad inside that he'd extinguish his own light? The laughter was the saving grace in an otherwise overcast life, clouded by addiction and bouts of depression.

Robin Williams is a testament to the complexity and fragility of our moods. Many scientists believe we are born with a baseline mood, much like a thermostat, that is set by our genetics and early environment. As Williams grew older, his mood was taken far from his baseline, due to a disease called Lewy body dementia, or LBD. First noted by Fritz Heinrich Lewy in 1912, Lewy bodies are abnormal protein aggregates in the brain that toss a monkey wrench into neuronal communication, causing victims to think and behave erratically.

More than a million people suffer from the disease, but we don't know why Lewy bodies form or how to get rid of them. As LBD progresses, patients can suffer from depression, insomnia, paranoia, and hallucinations. Robin's widow, Susan Williams, referred to LBD as "the terrorist inside my husband's brain."

The tragedy that befell Robin Williams demonstrates that our moods are firmly rooted in our biology; disease is just one example telling us that we're not in as much control of them as we like to think. Shazbot!

Where Your Feelings Come From

Emotions are difficult to define, but we sure know them when we feel them: love, hate, anger, joy, envy, empathy, to name a few. Our forebears used to think that spirits emanated from various organs to create these feelings within us. Love was thought to come from the heart, and anger from the yucky black bile in the spleen. But we soon realized that a damaged heart does not stop loving, and people who've had their spleen removed can still get mad.

But changes in emotional states do happen if the brain is injured or its biochemistry is altered. However magical they may feel to you, your emotions are purely biological in origin, generated by brain chemicals called neurotransmitters stimulating specific brain regions (note that some hormones also function as neurotransmitters). Emotions can also be felt when an electrical probe touches part of the brain, simulating the activity of neurotransmitters. As such, a large part of our emotional state is controlled at the genetic level, given that genes encode the enzymes that manufacture these neurotransmitters, the receptors they bind to, and the enzymes that break them down.

A wide variety of hormones and neurotransmitters are involved in creating emotions, so we'll just point out a few examples to drive home the point that biochemical signals govern our feelings. We've already met dopamine, the neurotransmitter that is released in response to what our body perceives to be important for survival or reproduction, like catching a fish or having sex.

Our brain did not evolve in this modern world of comfort, so it doesn't do a great job at distinguishing between trivial and nontrivial accomplishments. Therefore, dopamine will fire up whether you just get a great deal on your shoes, level up on a video game, or make it home without having to stop at a single red light. It is intertwined with our anticipation of reward, motivating us to pursue certain activities. As we learned in Chapter 4, malfunctions in dopamine signaling can drive people to risk-taking or addictive behaviors. Reduced dopamine is associated with a lack of motivation, procrastination, and a loss of confidence. Chronically low levels of dopamine can erase the ability to feel pleasure altogether, which is clinical depression; too much dopamine has been associated with aggression and psychiatric disorders, including schizophrenia and attention deficit disorder.

Although studies of dopamine have revealed a reward system in our brain, other research suggests that an "anti-reward" system is also present. Normally, the anti-reward system functions to bring us back down to Earth by releasing neurotransmitters to end the reward (all good things must come to an end!). However, in some people, the anti-reward system works too well and has been linked to depression and suicide.

We've also mentioned serotonin (also called 5-hydroxytryptamine or 5-HT), a neurotransmitter and hormone that is built from tryptophan, the amino acid that is erroneously implicated in putting you to sleep after eating too much turkey. Most people are familiar with serotonin because of common antidepressant drugs like Prozac (fluoxetine), which are thought to alleviate severe depression by keeping serotonin levels higher in the brain.

Even though serotonin is best known for its role in regulating mood and is present in the brain, most of it is found in our gut, where it promotes peristalsis (the process of moving things along from stomach to toilet). Serotonin is commonly associated with feelings of happiness and well-being, which may be linked to its other functions throughout the body; for example, many individuals with depression also suffer gastrointestinal problems, and vice versa. Recent studies have suggested that our microbiota are important for serotonin production, which most likely explains why our gut microbes are intimately connected to our gut feelings. Serotonin is also the precursor to melatonin, a hormone that regulates sleep, which has a huge influence on our mood.

Stress hormones are key in our fight-or-flight response. Of particular relevance to this chapter is cortisol, which is produced by the adrenal glands on top your kidneys in response to a perceived threat, such as catching a glimpse of what appears to be Stephen King's terrifying clown Pennywise out of the corner of your eye.

Cortisol keeps you alert and primed for action: It gets your heart pumping so blood can get oxygen to muscles faster, and helps release blood sugar to provide a burst of energy. At the same time, it puts other high-energy tasks, such as digestion and immunity, in time-out, to divert power to the threat. Ramping down the immune system also helps reduce inflammation in the event you are injured. These are useful responses if Pennywise is coming for you. But if it was just Ronald McDonald, your body needs a way to calm back down, so it can finish digesting breakfast and resume guard against pathogens.

That's where glucocorticoid receptors come into play. Glucocorticoid receptors, which are expressed in the brain and on immune cells, sop up the cortisol. Without the ability to dispose of the cortisol, a chronic stress response is experienced that can cause aggressive and paranoid behavior. Additionally, the immune system stays depressed, explaining why people who are stressed out all the time get sick more often.

Finally, sex hormones such as testosterone and estrogen are well-known mediators of our mood. Levels of these hormones wax and wane as we age, affecting our mood in subtle and dramatic ways. Estrogen deficiency can cause depression, fatigue, and memory lapses, while an excess can produce feelings of anxiety and irritability. Low testosterone (also known as "low-T") is associated with depression and fatigue; it can scatter focus and deflate the desire to have sex. Excess testosterone has been shown to make men more *(ahem)* cocky and blind to flaws in their reasoning, often leading to lots of bad decisions and reluctance to read daily intelligence briefings. And this is not just a man thing; women given oral testosterone think that they are invincible and ignore the input of others, too.

Testosterone is infamously associated with aggression, but pinning down whether it actually causes aggression across the board has been problematic. The hormone boosts optimism, confidence, and impulsivity—all things that may help one rise to a status challenge. But testosterone also spikes when performing acts of charity or generosity. These observations lead some scientists to argue that testosterone fuels behaviors that will advance social standing; whether the actions taken are aggressive or benevolent depends on the context of the situation.

The revelation that genes and biochemicals drive our feelings may seem mechanical and unsettling to some. But have no fear: Knowing how a car works doesn't make it any less fun to drive. Learning how emotions work at a molecular level does not rob you of the experiences they induce; despite all we learn here, you will still cry when you watch *This Is Us,* still laugh when you watch *The Office,* still cower in fear when you watch *Halloween,* and still get angry when you bite into a chocolate chip cookie and realize it's oatmeal raisin.

Demystifying our feelings does not in any way diminish their impact on the human body and our behavior, but it does remind us of our rare gift: the ability to control our emotions, rather than be controlled by them. After all, emotions are just one input our mind uses to construct a facsimile of the world, and the information they parlay is not infallible. (Remember, sometimes Pennywise is just Ronald McDonald.) Acknowledging that emotions are biochemical signals that may be right or wrong should motivate us to look upon these impulses through the lens of reason, rather than being driven purely by thoughtless instincts. And although emotions are important voices in the democracy of the mind, we should not allow any one of them to become a dictator.

Emotional responses are fast and furious, but they pave the way to a mood, which develops when there is a sustained emotional experience. For example, people like Charlie Brown, who experience repeated misfortune, often gradually develop a more anxious baseline, meaning they remain anxious even when things are going well. A mood disorder develops when a person's baseline reaches a point when an emotion feels so persistent that it seems inescapable. With all of the aforementioned signals working on the brain, it is easy to appreciate how maintaining mood is a very delicate balancing act for the body. It doesn't take much to gum up the works.

Why You Seem Down in the Dumps All the Time

Everyone gets the blues from time to time, but clinical depression goes way beyond a normal funk. It sets in when someone is suffering from unremitting sadness to the point of anhedonia, the loss of interest in activities that bring joy. Underscoring the severity of the disorder, individuals with depression can't even derive pleasure from food and sex (or if they do, the elation is fleeting and they quickly retreat back into misery).

The World Health Organization estimates that more than 300 million people suffer from depression worldwide, with nearly 800,000 committing suicide each year. If we could identify the genes that are linked to depression—or, conversely, which genes are linked to happiness—it should lead to new drugs that alter mood in positive ways. The search for factors that change the color of our mood ring has been like a genetic gold rush. But finding that pot of gold at the end of the rainbow has proven more challenging than anticipated.

After decades of combing through genome sequences, it has become apparent that many diseases and behaviors are too complex to be explained by one gene. There are exceptions: For example, certain variations in genes for Huntington's disease or cystic fibrosis will virtually guarantee that you'll develop those conditions. But there is no single gene for depression.

There is strong evidence, however, that depression has a genetic component. Depression is known to be heritable: In other words, it runs in families. Studies of identical and fraternal twins have revealed that depression is about 37 percent heritable. That finding confirms a genetic component to the disorder, but also means 63 percent of the cause comes from elsewhere (for example, the environment). It is also apparent that depression is polygenic, caused by multiple genes.

Hunting for genes that cause something like depression is like trying to find a needle in a haystack. The strategy is simple: Compare the DNA sequence from lots of depressed people with the DNA sequence of lots of people who are not depressed. Seems easy enough—and as it happens, we have an amazing technique called genome-wide association studies (GWAS) that employs computers to compare DNA sequences from thousands of different people. The problem is that a ton of differences between the DNA sequences of any two people have nothing to do with depression. We need to find a way to pull out the signal (genetic differences associated with depression) from the noise (genetic differences not associated with depression).

Despite the promise of GWAS, the first several studies turned up bubkes. How depressing that all this fancy technology can't find the genes for depression. But in 2015 geneticist Jonathan Flint at the University of Oxford got an idea how to better tune GWAS to isolate the signal from the noise. Flint appreciated that depression is a complex disorder that varies in its magnitude; some cases of depression are mild and sporadic, whereas others are persistent and devastating. He decided to focus his genetic comparisons only on people who were suffering from recurrent, severe depression. To minimize the noise of unrelated genetic variation further, he only studied women of Han Chinese ethnicity.

After examining the DNA from about 5,300 Chinese women with severe depression and comparing it with about 5,300 controls (Chinese women without depression), Flint and his team identified two gene variants linked to depression. One variant is in a gene called LHPP, which encodes an enzyme whose function remains to be characterized. The other variant is in SIRT1, an enzyme involved in crucial cellular organelles called mitochondria. Mitochondria are the "powerhouses" of the cell that generate energy storage molecules called adenosine triphosphate (ATP). Mutations in SIRT1 might explain why people with depression often suffer from lethargy, but scientists have much more work to do to understand how these gene variants may influence depression. It should also be noted that these two genetic variations do not appear with high frequency in people of European descent, so depression among other ethnicities may arise from different genes.

Taking a page from Flint's playbook, a 2017 study looked exclusively at people with major depression within the limited gene pool of an isolated village in the Netherlands. Researchers found a new variation in NKPD1, a gene that was subsequently observed to also contain variants in people with depression beyond this village. The gene defect may lead to altered levels of sphingolipids, which serve as a signaling molecule in the brain, among other things. Intriguingly, two known antidepressants inhibit sphingolipid synthesis.

Another way to pull out the signal from the noise is to vastly increase the sample size—the more people with depression who are profiled, the more confident we can be about the differences that come up consistently. The personal genome service company 23andMe is stepping up to study complex disorders like depression. In comparing their vast library of consumer DNA samples, they have uncovered 15 new DNA regions that may be linked to depression risk. Some of the genes involved seem to make sense, as one is known to function in learning and memory, and another is involved in neuron growth.

Historically, a few other genes have received a lot of attention as being potentially linked to depression. A great deal of work has implicated serotonin in depression, and several high-profile antidepressants like Prozac and Zoloft are believed to target the serotonin system. Genes encoding components of the serotonin system, including serotonin transporters (5HTT/SLC6A4) and receptors (HTR2A) have been linked to depression symptoms. There is also evidence that brain-derived neurotrophic factor (BDNF) plays an important role in depression. BDNF is important for neuron development in the brain, and decreased levels are seen in animals subjected to stress and in humans with mood disorders.

In parallel to the hunt for genes underpinning depression, a great deal of work has convincingly shown, unsurprisingly to many, that stressful life events are key components to the development of depression. Some of the biggies are loneliness, unemployment, and relationship stressors. But topping the chart is childhood abuse or neglect. A landmark 2003 study by psychologist Avshalom Caspi at King's College showed just how important gene-environment interactions can be. Caspi and his colleagues revealed that one of the serotonin transporter gene variants is more tightly linked to depression if the individual suffered from adverse life events. This is an important result that helps explain why the gene hunting has been so difficult: Not everyone with a gene variant associated with depression will actually develop depression. Investigating exactly how the environment can tweak genes toward depression is the new frontier in mood research.

How Your Childhood Shapes Your Mood

In the right genetic context, adverse childhood experiences (ACEs) can set up an individual for a lifetime of susceptibility to depression triggers. Science is uncovering numerous ways this can happen, and a common thread is that ACEs reprogram genes responsible for wiring brains, causing them to be more sensitive to stress. From a 2017 study, neuroscientist Catherine Peña at the Icahn School of Medicine at Mount Sinai revealed that when baby mice were stressed shortly after birth, they had reduced levels of a transcription factor called OTX2. The job of a transcription factor is to turn certain gene networks on or off; these factors complete their jobs at discrete times during development or in response to certain environmental conditions.

In Peña's study, the stress-induced loss of OTX2 early in life had serious and irreversible effects on how the mouse brain developed, leaving it prone to depression. The interesting part is that OTX2 levels normalized as the mice grew older. But the brain damage had been done, and once they experienced stress as adults, the mice lapsed into depression. If they did not experience that second hit of stress as an adult, they remained normal. Such experiments teach us that brain development during infancy and childhood is critical in setting people up to handle stress better later in life.

It has long been known that children who were abused grow up with an increased risk of physical health problems like type 2 diabetes and heart disease, as well as psychological problems including depression, drug addiction, and suicide. The aftermath of childhood adversity can plague the victims long after the trauma has ended—and, in some cases, even if they were removed from the abusive situation and raised in a nurturing environment.

Epigenetics is providing a biological basis for the ghosts that follow childhood trauma. A pioneering 2004 study by neurobiologist Michael Meaney and geneticist Moshe Szyf at McGill University showed that rat pups of inattentive mothers grow up to be very anxious, with more DNA methylation at a gene called NR3C1. NR3C1 encodes a glucocorticoid receptor that cleans up the stress hormone cortisol. Because the gene is highly methylated, less of the glucocorticoid receptor is made, and the stress hormones don't get vacuumed up. This chronic pollution of stress hormones causes the body to suffer mentally and physically. In the pups of attentive mothers, the gene for the glucocorticoid receptor is rarely methylated, so they grow up to handle stress normally.

The same situation appears to hold true in humans. Genetic analysis of children who were abused and later became suicidal also showed increased DNA methylation at their NR3C1 gene. DNA extracted from blood samples provided by children of abuse also shows higher methylation at the NR3C1 gene. Extensive differences in DNA methylation patterns across thousands of genes are also seen among children raised in orphanages, compared with those raised by their biological parents. The epigenetic changes seen in the orphans were concentrated in genes regulating the brain and immune system.

Since Meaney's work, other researchers have discovered additional epigenetic changes occurring in rodents or humans subjected to childhood adversity, many of which are detected at genes associated with brain function or stress management. These groundbreaking studies reveal why many abused and neglected children can't "get over it," as outsiders sometimes naively wonder. ACEs don't just get under the skin; they get into victims' DNA, scarring their genetic code in ways we are just beginning to understand. Whether these DNA scars can be reversed is a subject of intense research, as is what makes some children exposed to ACEs more resilient.

Even more disturbing is evidence that some of the epigenetic changes that occur during ACEs can be inherited, potentially perpetuating a parade of crummy parenting through generations. Meaney's group found that pups born of inattentive rat mothers have increased DNA methylation at genes encoding estrogen receptors, which led to a decreased level of these hormone receptors in adulthood. Without reduced ability to bind estrogen, these females do not receive a loud and clear signal to be nurturing. In other words, the DNA of the neglected rats was programmed in a way that made them neglectful mothers, just like their own mothers.

Studies have also demonstrated that children's socioeconomic environment can affect their epigenomes, providing biological justification to urgently invest in impoverished neighborhoods and schools. A 2017 study led by neuroscientist Douglas Williamson at Duke University found that adolescents growing up in homes with lower socioeconomic status had higher DNA methylation at the gene SLC6A4. As a consequence of lower levels of this serotonin receptor, their brains undergo developmental changes resulting in a hyperactive amygdala, the brain region associated with our fear response and how we respond to threats. Growing up in poverty induced epigenetic changes that led to an amygdala stuck in overdrive, most likely explaining why these teens report symptoms of depression later in life.

In addition to stressful events in the environment, a person's culture may also affect how their genes have evolved and function. This relatively new field, called dual inheritance theory, examines how genes and culture influence each other. The genetic variations of the serotonin transporter mentioned previously (5HTT) appear to have different effects on mood, depending on whether the carrier lives in a culture that values individualism (such as North America) or collectivism (for example, East Asia). The variant of 5HTT that has been linked to depression is highly prevalent in East Asia compared with the United States, and yet more people in the United States develop severe depression.

What might explain this discrepancy? Other genes or differences in diagnostics, possibly. But some researchers attribute these differences to culture. It has been suggested that the mutant 5HTT does not necessarily predispose people to depression, but causes them to be more sensitive to both positive and negative experiences, particularly social ones. With heightened sensitivity to social interactions, carriers of the 5HTT who live in collectivist societies that offer more social support, rather than a "go at it on your own" attitude, may be better shielded from depression. Some studies back this idea. If children carrying the depression-prone 5HTT gene have a mentor in their life, they are significantly less susceptible to developing depression; on the other hand, children who carry the 5HTT variant and who are maltreated and have no positive support system have the highest depression ratings.

In the case of depression, genes are clearly important. But so is the environment, especially during childhood. Providing children with a strong social support system shows promise in minimizing the development of severe depression in adulthood, even for those who are genetically predisposed.

How Your Gut Affects Your Mood

Before the year 2000, if you were to tell scientists that lowly gut bacteria could affect someone's mind, they probably would have laughed their safety goggles off. But the creation of germ-free mice changed everything. As you'll recall, mice kept in sterile conditions and lacking microbiota are far from normal. In addition to the weight problems we discussed in Chapter 3, some strains of germ-free mice are as neurotic as George Costanza, and their stress hormone levels are through the roof.

One of the ways scientists can tell whether a mouse is anxious is to put it into an elevated "plus maze." The maze is a big plus sign with two arms that are open to the lab and two arms that have private enclosures. Anxious, germ-free mice gravitate to the enclosed areas and are reluctant to explore the open arms, as normal mice tend to do. Naturally, scientists wondered what would happen if they put bacteria into these nervous germ-free mice.

Transplanting *E. coli* into the germ-free mice didn't restore normal behavior, but giving them bacteria called *Bifidobacterium infantis* did. The finding shows that bacteria can restore normal behavior. But they can't be any old bacterial species; they have to be a specific kind. These discoveries were completely unexpected, leading to the idea that bacteria in our gut may do more than help digest food. They could be influencing behavior, personality, and mood.

In 2011, neurobiologist John Cryan at University College Cork in Ireland conducted an exciting experiment that has sparked new interest in the potential of probiotics. In the study, mice fed bacteria called *Lactobacillus rhamnosus* (a widely used strain in commercial probiotics) displayed lower stress hormones, reduced anxiety, and decreased depression-like behavior. Because mice aren't inclined to lie down on a psychiatrist's couch and discuss their problems, researchers commonly use a life-threatening challenge, like a swim test, to gauge depression. Plop the mouse into a bath and see if he tries to swim (normal) or if he says, "Forget this" and gives up (depressed). (The researchers, of course, rescue the little critters from drowning.)

Multiple lines of evidence show that our gut bacteria may exert a level of control over our minds and moods, too. The idea started percolating in 2000, after a flood contaminated the water supply and left many of the residents of Walkerton, Canada, reeling from bacterial dysentery. After the acute gastrointestinal ordeal was over, many folks in the town developed irritable bowel syndrome. Years later, scientists also recorded a significant spike in depression among those who fell ill during the Walkerton flood, believed to be the result of an imbalance in their gut bacteria caused by the earlier intestinal infection. Could tiny microbes teeming in our bowels really contribute to serious mood disorders like depression or anxiety?

Other researchers have subsequently found that people with depression harbor a different microbiota than people without a mood disorder. But a key question remained: Do intestinal bacteria cause changes in mood, or do changes in mood alter intestinal bacteria? In 2016, Cryan and his colleagues addressed this question of causation by testing whether the blues were contagious, and if they could be transmitted by someone's bacteria. Astonishingly, germ-free rats that received intestinal bacteria from human patients with depression developed symptoms of depression, displaying anxious behavior and showing no interest in sweet treats. Germ-free rats colonized with bacteria from people who were not depressed were fine.

How can itty-bitty bacteria in our gut have such a big effect on our head? Studies have shown that our microbiota generate a variety of neurotransmitters and hormones that could directly affect the way we think, feel, and act. In Cryan's 2011 study, his team uncovered one way bacteria in the gut could get word to the brain. When they severed the vagus nerve, which is the major neurological conduit connecting the gut and the brain, the neurochemical and behavioral effects that implanted bacteria bestowed were abolished. In addition to being vagus nerve whisperers, gut bacteria can also influence the brain indirectly through interactions with our immune system, which assigns many beat reporters to wander the intestines and send news to the brain.

You're probably wondering by now whether probiotics can alter our gut bacteria in ways that alleviate anxiety or depression. Although data are still scant and preliminary, one study reported that women who consumed a probiotic yogurt for one month showed alterations in brain regions that control the processing of emotion and sensation. These women showed dampened reactions to scared and angry faces, suggesting that yogurt might help you stomach *The Walking Dead*.

Another study from 2011 showed that stress hormones were lowered in folks taking probiotics, and a more recent trial also supports the use of probiotic supplements to help reduce negative thoughts associated with sad mood.

But complicating the interpretation of these studies is the fact that all probiotics are not created equal; they vary in which microbes are used and how many of them are present (measured in colony-forming units, or CFUs). Because microbes in probiotics can be affected by someone's indigenous microbiota, their diet, and their genetics, it is likely that these treatments will affect people in disparate ways. Although generally considered safe, some studies have linked probiotics with adverse effects such as bloating and brain fog. In short, much more work needs to be done to prove the efficacy of probiotics.

Why There Are Grumpy Old Men

"Get off my lawn!" many an old man has yelled, most famously the rifle-toting Walt Kowalski played by Clint Eastwood in the 2008 film *Gran Torino*. What's with grumpy old men, anyway? Don't they have bigger fish to fry? Why are they always so uptight, ranting and raving about how awful the world has become?

In the thick of middle age, I now have much more sympathy for the stereotypical grumpy old man. Usually retired, with a collapsing circle of friends and kids who have moved on, seniors sometimes feel like they've lost their utility in the world. Technology is advancing, and yet their mental hardware is in decline. On top of all that is the stark realization that they are in the twilight of life, possibly spinning their "Cat's in the Cradle" 45 rpm record for the last time. These are understandable reasons for crankiness. But not every senior becomes Oscar the Grouch.

Science is on the case. Researchers have named the phenomenon irritable male syndrome; on average, it is most likely to begin at age 70. This figure corresponds with the age when testosterone levels plummet. Recall that low testosterone is associated with irritability, difficulty concentrating, and negative mood, providing a biochemical explanation for the proverbial grumpy old man. In some men, testosterone levels do not diminish as rapidly, possibly helping them to stay chipper for another decade or two. Other health problems such as kidney disease or diabetes can accelerate the decline in testosterone.

Microbiologist Marcus Claesson at University College Cork in Ireland has been looking into another change that accompanies old age: the microbiota. Claesson has found that the composition of intestinal bacteria in elderly people is different compared to young whippersnappers. Interestingly, older people tend to have fewer of the bacterial species that have been suggested to be helpful in dealing with stress. The aged microbial landscape, which in some cases was linked to changes in diet that accompany elderly care, may also contribute to the increase in pro-inflammatory immune signals and frailty generally seen in geriatrics.

Finally, as we get older, our medicine cabinet begins to resemble a pharmacy, and each one of these medications has the potential to alter mood. Some of these medicines, chiefly antibiotics, can alter the intestinal microbiota as well.

So show a little compassion. Use the sidewalks. And maybe treat a grumpy old man to some yogurt.

Why You Get the Winter Blues

Songs have been written about the summertime blues, but not so many about the very real wintertime blues. Up to 6 percent of people in the United States suffer from seasonal affective disorder, aptly abbreviated SAD. People with SAD experience anything from mild anxiety to serious doom and gloom during the winter months. The condition is more common in northern areas where the days are even shorter during winter. Why does this happen?

Like all other animals, our body is equipped with an internal biological clock that adjusts our metabolism to fit the demands of the day and to sleep soundly at night. The waning of daylight is a primal cue that literally tells our body that it is time to sleep. When the cells in our retinas stop sensing light, they signal our brain to make melatonin, the sleep hormone that serves as a biochemical lullaby. When that morning sunshine hits our eyelids, which are transparent enough to let changes in brightness through, our brain puts the brakes on melatonin production so we can be awake for the day. Or at least awake enough to make some coffee.

When we use artificial lights at night or stare at bright screens before bed, our brain gets confused. Thinking that it is still daytime, our brain doesn't make our melatonin nightcap, which makes it harder to catch some zzz's when we finally turn the lights off. A similar situation arises with people who have SAD: Their body's production of melatonin is out of sync with the sun, which is made worse when there is less sunlight during the day.

Several gene variants have been linked to SAD, including genes that regulate our biological clock. Another SAD gene variant encodes a receptor for serotonin, which is of interest because it's the precursor molecule from which melatonin is made. Some people with SAD have a variant OPN4 gene, which works in the eye to detect light and signal the brain to crank out melatonin.

A recent study conducted in 2016 by neurologists Ying-Hui Fu and Louis Ptáček at the University of California, San Francisco, discovered yet another gene that underscores the importance of light and sleep in setting our mood. Variants in a gene called PERIOD3 were found in people who had both SAD and familial advanced sleep phase (FASP) disorder. People with FASP have a biological clock that runs fast, meaning they feel the need to sleep very early, say 7 p.m., and then wake up around 4 a.m. When the researchers put this variant form of PERIOD3 into mice, they behaved normally when day and night were of equal duration. But when exposed to shorter periods of daylight, as patients with SAD would be in the winter, these mutant mice gave up too easily when put into mildly stressful conditions: symptoms that are consistent with depression.

SAD is usually treated with bright light therapy in the morning to stop melatonin production, and with melatonin supplements in the evening to induce sleep. Resetting the biological clock in patients with SAD helps alleviate the adverse moods they experience due to poor sleep. The broader lesson that we learn from patients with SAD is not to underestimate the value of a good night's sleep on our well-being.

How Carbs Can Make You Giddy

Strange reports have surfaced in recent years describing individuals who appear to be completely drunk without taking a sip of alcohol. A senior citizen starts acting three sheets to the wind after having his morning bagel. A woman from upstate New York is charged for driving under the influence (DUI), though she swears she is innocent and had nothing to drink. A three-year-old girl starts acting intoxicated after drinking fruit punch. What is going on here? Are these people playing us? Are they compulsive liars?

The woman who was charged with DUI was put to the test in a skeptical courtroom. She was monitored for a 12-hour period, taking a Breathalyzer test every few hours. Despite having no booze whatsoever, her blood alcohol content rose steadily throughout the day, reaching four times the legal limit by the end of the period. The judge decided to dismiss the charges. She was not consuming alcohol, but her body was making it somehow.

What turned out to be happening here is one of the most bizarre examples of microbes affecting mood and behavior. When certain intestinal yeasts overgrow, they can produce alcohol from ingested carbohydrates in a condition called auto-brewery syndrome or gut fermentation syndrome. People with this syndrome can feel inebriated when all they had was a big plate of pasta. Yeasts are a type of fungi, and those that have been linked to auto-brewery syndrome include *Candida* species and *Saccharomyces cerevisiae* (particularly ironic given its common name: brewer's yeast). Yes, that is the same yeast your hipster friend uses to make his craft beer, so people harboring excessive *Saccharomyces cerevisiae* in their intestines develop a literal "beer gut."

It is not known why these yeasts gain a foothold in the gut of these people. One documented case suggests that prolonged antibiotics, which affect bacteria but not fungi, can create an environment in the intestine that favors growth of the yeast. With the bacteria depleted, there is less competition for nutrients, so the yeast can party on. Some researchers have argued that overgrowth of fungi is not to blame; rather, these people may have genetic defects that prevent the liver from metabolizing the minute, normal levels of alcohol that may ferment in the gut.

Although the extreme cases get the most attention, imagine how fluctuations in the yeasts present in your gut might modestly lighten your mood or impair your judgment by producing smaller quantities of alcohol. But don't plan on using this strategy to defend your questionable behaviors anytime soon. Auto-brewery syndrome is exceedingly rare and can be remedied with dietary changes, antifungal drugs (which target the yeast infection), and probiotics (which attempt to replenish the gut with healthy bacteria).

Why There Are Shiny Happy People

Although it is generally accepted that depression is rooted in biology, many people still believe that happiness is all about your state of mind. They contend that if you have the right attitude, you can turn any frown upside down. Historically, achieving a state of worry-free nirvana has been a quest left to philosophers, theologians, and Bobby McFerrin. But scientists want to be invited to this happy hour, too. To paraphrase Clark W. Griswold from *National Lampoon's Vacation,* maybe science can find the key to happiness and have us all whistling "Zip-a-Dee-Doo-Dah" out of our posteriors.

Naturally, the hunt for genes associated with increased happiness is on. But where to look? Bristol University economics professor Eugenio Proto, in research conducted while he was at the University of Warwick in the U.K., figured a good place to start would be to analyze DNA from people living in Denmark and other Scandinavian countries—nations that regularly top world happiness rankings (which may also explain their penchant for ABBA). Of the 30 countries examined, Denmark and the Netherlands have the lowest percentage of people carrying a serotonin receptor gene implicated in depression. It seems there might indeed be something about Danish DNA that makes them happier than most. But the free college and health care probably help, too. As a testament that money isn't everything, the United States consistently ranks in the teens (dropping from 14 to 18 in 2018), well below comparatively wealthy nations.

Other groups have identified genes that might allow some people to have a lighter baseline mood. In 2016 geneticists found a strong correlation between a nation's happiness and a variant of the gene encoding the enzyme fatty acid amide hydrolase (FAAH, mentioned in Chapter 4). This variation interferes with the breakdown of the neurotransmitter anandamide, the "bliss molecule" mentioned in Chapter 4 (recall that it binds the same receptors that the THC in marijuana binds). So perpetually happy people may naturally have more anandamide in their system; this not only augments sensory pleasure but also has painkiller properties. As with other genetic variations, though, the environment interplays with DNA and makes the outcome less predictable. Case in point: The researchers stated that some countries, such as Russia and those in Eastern Europe, also have citizens who possess the FAAH "happiness" variant and yet do not consider themselves to be very happy.

In one of the largest genetics studies performed to date, on nearly 300,000 people, biological psychologist Meike Bartels at Vrije Universiteit Amsterdam discovered three new genetic variants commonly present in happy people. These gene variants are expressed in the nervous system, adrenal glands, and pancreas, but more work needs to be done to understand how they might relate to mood.

Although the research is young, it appears that genes influence our baseline happiness level in the same conceptual way that they influence our cholesterol levels. Additional evidence that we have a "set point" mood like a thermostat comes from a classic psychology study performed in the late 1970s that examined happiness levels after major life events.

Who do you think would be happier or sadder—a lottery winner or someone who became paralyzed as the result of an accident? Defying intuition, it turns out that several months afterward, lottery winners were not much happier than before their win, and the people with paralysis were not much sadder than before the accident. Although everyone would prefer to win the lottery, what we don't factor in is the stubborn ability we have to adapt to our situation—whether good fortune or misfortune. In other words, after the initial high or low, our mind normalizes and we tend to return to the baseline mood established by our genes, fetal programming, and early childhood environment.

Why Happiness Might Not Be All It's Cracked Up to Be

For some people, it seems like nothing ever gets them down. They are relentlessly happy, gleefully batting away any obstacle like an enthusiastic kid playing Whac-A-Mole. Although the escalation of sadness to depression is pathological, could there be something wrong with being happy 24/7? John Mellencamp seems to think so. He's been quoted as saying, "I don't think we're put on this earth to live happy lives. I think we're put here to challenge ourselves physically, emotionally, intellectually." As we'll soon see, science suggests Mr. Mellencamp has a point.

The problem with being too content was the basis for *Rocky III*. Our beloved boxing hero, Rocky Balboa, got too cozy as the reigning heavyweight champion. After defeating Apollo Creed, he was living large with Adrian, enjoying fancy cars, a home theater, and a robot. He didn't take the challenge of Clubber Lang seriously, and paid the price by losing the fight and the championship title. The moral of the story is that to win and stay a winner, you need the eye of the tiger. Feeling bad can restore the hunger and drive needed to achieve your goals. If Rocky had just shrugged off the loss and gone back to his mansion to roll in his dough, it would have put the franchise down for the count. But he fed off his negative feelings, using them as fuel to get through the training montage and win again.

University of Colorado Boulder psychologist June Gruber is one of the Debbie Downers skeptical of happiness. In her article "A Dark Side of Happiness?", written while she was at Yale University, Gruber outlines several circumstances when happiness is not appropriate, and possibly even detrimental. For example, our emotional state is important for communicating our status to others. If we are happy all the time, others can't tell when we need their help and support. Similarly, we're inviting trouble if we joke at a funeral, fail to show remorse after hurting someone, or sport a goofy grin when the boss is chastising us. Continually happy people could be viewed as cocky, aloof, unable to take anything seriously (or even as dumb and naive). As we learned from Rocky, people who are comfortable and content may underestimate threats and lack the drive needed to solve problems. Overly happy people have also been shown to be less adept at formulating persuasive arguments and tend to be more gullible.

A body of literature suggests that the pursuit of happiness, our inalienable right, can actually have a negative result on well-being. A 2018 study by Sam Maglio at the University of Toronto Scarborough, Canada, showed that people who are obsessed with trying to be happy perceive time differently from folks who appreciate what they already have. Those still chasing happiness often feel rushed and short on time, feelings that hinder happiness. According to the study, people who feel like they never have enough time in the day tend to refrain from leisure activities that would provide them the very happiness they seek. In addition, busy bees usually don't feel like they have the time to help others or do volunteer work, both of which are activities that make people happy.

Dismissing negative emotions may undermine the body's ability to calibrate itself correctly to navigate the hurdles standing in the path to a goal. In other words, we need the biochemicals associated with negative feelings because they help us fight through obstacles. If true, it would seem science has validated Mr. Mellencamp's assertion that we're not supposed to be living happy lives but challenging ones. Indeed, evolution has selected for coping mechanisms that enable us to thrive in less than hospitable environments.

Real-world examples for these ideas can be witnessed in children who are born with Williams syndrome, which results from a genetic defect that eliminates nearly 30 genes. Children with this condition suffer a range of cognitive and physical problems. But one of the unexpected features of the condition is unbridled happiness. These endearing individuals are unusually sociable, trusting, and polite, often to a fault. For example, many of these children instantly and openly love everyone. They do not fear strangers, and seem incapable of distrust. Unfortunately, this makes them easy prey for scammers, bullies, and child predators.

In addition, people with manic depression can get into serious trouble during their mania phase, which is characterized by an intensely positive mood. Persistent happiness can make people blind to obvious dangers, putting them in the most precarious situations. Manic behavior has included giving away our life's savings or speeding off to meet a stranger for an adulterous love affair.

None of this is to say that we shouldn't ever be happy, of course. But happiness is like bacon: Too much can be unhealthy. Adversity is an inevitable part of life, and denying the unpleasant emotions that accompany difficulties deprives us of the physiological tools we need to get through a challenge.

Clap Along if You Know What Happiness Is to You

For those of us without mood disorders who are just trying to turn up the dial on our happiness, what do experts have to say in the way of advice? One could spend a lifetime reading all there is on the subject of happiness. But after perusing a few key books, some common threads emerge.

The spiritual guru Ram Dass wrote a famous book in 1971 called *Be Here Now,* which inspired the George Harrison song of the same title. Dass promoted living in the moment and not obsessing about the future. More recently, the Harvard psychologist Daniel Gilbert expounded on this concept in his 2006 book *Stumbling on Happiness.* Gilbert explained how our uncertain future creates dissonance for our brain, which is a major control freak (more on this in Chapter 8). Moreover, we usually do a bad job imagining what will make us happy down the road, because what pleases us now is not guaranteed to thrill us then. In short, the future is a great source of anxiety that hinders happiness.

These ideas echo the observations in the philosopher Bertrand Russell's 1930 book *The Conquest of Happiness.* Russell instructed people to "become emancipated from the empire of worry" by realizing the unimportance of most matters in the grand scheme of things. The Universe doesn't care which outfit you decide to wear or how many material things you can hoard. Russell claims to be closer to fine when he diminishes his preoccupation with himself.

Similar sentiments are proffered by ethicist Peter Singer, who advocates that the greatest happiness is not found by fulfilling narrow, self-centered goals but by taking on the challenge of making the world a better place for as many as possible.

Evolution may have begun with selfish genes, but we've evolved a remarkable capacity for altruism. Our natural tendency to cooperate emerged with kin selection, the instinctive drive to support those who are genetically similar to us, such as our family. Over time, we found that it is also helpful to support other members of our community. Singer argues that we now need to expand this circle of moral concern globally to all members of the human race. If you don't trust crusty academics, comparable insight can be found in the comics, such as when The Ancient One explains to Dr. Strange: "Arrogance and fear still keep you from learning the simplest and most significant lesson of all . . . It's not about you."

In agreement with these philosophical ideas, scientific research has shown repeatedly that the happiest people are those who focus on others, rather than themselves. Brain imaging studies have revealed that acts of altruism light up the same reward center in the brain that food and sex activate. Helping others infuses your life with a sense of purpose and meaning; it provides instant gratification and near-euphoric satisfaction. It also helps you see a grander view of the world and the diverse people who share it with you. Albert Einstein was onto this idea when he said, "Strive not to be a success, but rather to be of value." Helping others is not only the humane thing to do, but it is the key to happiness.

Awakenings

In 1969, famed neuroscientist Oliver Sacks made the astonishing discovery that L-dopa, the precursor to several neurotransmitters, including dopamine, could revive patients who had been in a catatonic state for decades. This single chemical seemingly flipped a switch back to the "on" position in some of these patients after years of being unable to speak or move. Unfortunately, their resurrection was short-lived; some built up a tolerance to the drug or suffered dangerous side effects. The incident underscores a cold reality that we must be courageous enough to face: All our thoughts, emotions, and feelings are biochemical in nature. Our moods arise from our neurophysiology, not from a mystical spirit within. Anything that can alter our brain, even in minor ways, can alter our mood.

Sacks's account of this extraordinary event is described in his book *Awakenings*, which was made into a movie starring Robin Williams as the brilliant and unassuming doctor. Unbeknownst to Williams, malicious proteins that would claim his life 24 years later were slowly accumulating in his brain.

As with other celebrity suicides, the days following Robin Williams's death were rife with commentators expressing bewilderment at how he could be so selfish or such a coward. Such comments are not merely ignorant but are reprehensibly cruel to grieving loved ones. Perhaps we will come to accept that hidden forces like our genes, epigenetic programming, and microbiota largely and unconsciously shape our sunny or not-so-sunny dispositions. Mood disorders are a very real health problem; both excessive sadness and excessive happiness are detrimental to our normal operations. Berating someone to snap out of it is like yelling at a blind person to see. A more helpful approach is to provide support and encourage professional help.

The fact is that science has led to a human awakening: Our baseline mood is predetermined to a large degree early in life by factors beyond our control. We also don't have much say in the alterations that appear in our mood as we age. Some would say these revelations are a rude awakening: We prefer to be in control, and we want the power to change our mood. We can get there, but not without embracing the truth. As we continue to better understand the biological basis for mood problems, a new battery of treatments will become available.

» CHAPTER SIX «

MEET YOUR DEMONS

We never lose our demons.
We only learn to live above them.

—The Ancient One to Mordo, *Doctor Strange*

I spent all my life on the East Coast until I ventured out to the Midwest to pursue postdoctoral studies. Shortly after moving from Philadelphia to Indianapolis, I was driving home on the interstate one night when a guy in a large pickup truck abruptly cut me off and nearly caused a serious accident. Having been trained as an East Coast driver, I laid on the horn for about half a mile and showcased one of my digits to him numerous times.

As it happens, the guy takes the same exit as mine and before I know it, I'm trapped behind him as the traffic light turns red and cars line up behind and beside me. Now I notice his NRA bumper sticker and gun rack. His truck also has mud flaps with Yosemite Sam pointing guns at me, saying "Back off." The driver's door opens. My eyes widen. My heart races. I start to sweat. My anger quickly dissolves into an intense fear.

Next, an impossibly tall and rugged man steps out of his truck and tugs on his belt buckle, which is roughly the size of Texas. He marches toward my car and taps on my window with his skull ring. Naturally, I look straight ahead and pretend nothing is happening. He knocks on my window again and calls, "Hey!"

Pinching my legs together so not to pee my pants, I turn and look at him through my car window. He runs his hand over his handlebar mustache and motions for me to roll it down. *No way, dude.* Soon, he starts shouting at me, but not the words I expected to hear.

"I am so sorry! I didn't mean to cut you off back there, I just didn't see ya." His voice was shaking, and he seemed as scared as I was. I cracked the window open, told him it was OK, and apologized for overreacting and blaring the horn like an idiot. He tipped his hat to me and ran back into his truck as the light turned green.

I was fortunate this fellow was thoughtful and levelheaded; we read of many instances where road rage leads to a tragic outcome. But what determines whether some people let loose their demons? Are some people naturally passive while others are more aggressive? Could evil spirits be the source of demonic behavior? Before we answer these questions, it will be helpful to take a closer look at the biology of fear.

Why You Get Scared

No matter how big and strong we are, fear happens to us all (even the Man of Steel is afraid of kryptonite). But as unpleasant a sensation as fear may be, it's a necessary evil that evolved to protect us. There is a strong survival advantage for genes that build nervous systems quick to respond to perceived threats. The faster a survival machine can address a threat, the more likely it will remain alive to reproduce, passing this robust fear response to the next generation.

Fear is an autonomic response, meaning it happens without our thinking about it. This is why you jump when your friends shout "Surprise!" at your surprise birthday party. You don't stand in the doorway mulling it over and then decide to jump.

Rapid biochemical changes take place in our body the instant we become frightened. In response to a stressful stimulus—an unexpected noise or ominous shadow appearing behind you—your brain goes into red alert and signals the release of stress hormones. Epinephrine (adrenaline) and norepinephrine are released to increase breathing, heart rate, blood pressure, and the breakdown of stored sugars for energy. They also dilate your pupils for improved vision and switch off digestion to divert energy to the threat. Cortisol is also released, which increases blood sugar and suppresses the immune system, both of which provide extra energy to deal with the threat. All of these compulsory actions are critical for your fight-or-flight response.

How does the body know what to fear? Everyone jumps when startled, but not everyone is spooked by a New Orleans Ghost Tour. Things that animals are programmed to fear without being taught are called innate fears. For example, mice are born with an innate fear of cats. Cats are born with an innate fear of dogs (and cucumbers, according to YouTube). Loud, unexpected noises and the fear of falling are the only two known innate fears in humans; the others must be learned (although we learn to fear some things quickly and easily, such as spiders, snakes, and Dick Cheney). People who have been taught to believe in the supernatural are more likely to fear ghosts than nonbelievers.

Whether fears are innate or learned, genetic variation can partly explain why some people freak out and others chill out. In Chapter 5, we discussed a variant in a gene called FAAH that leads to more anandamide, the bliss molecule that makes people less anxious. These individuals also have enhanced fear extinction, or the ability to become unafraid of something faster than others. Mutations in receptors for the inhibitory neurotransmitter GABA (introduced in Chapter 4) have also been linked to fear. Damaged GABA receptors block the brain from receiving this "calming" inhibitory neurotransmitter, providing a likely explanation why some people are more frightful and anxious. Mice deficient in GABA receptors are bigger chickens than normal mice. Mutations in GABA receptors have also been found in people suffering from panic disorder, a condition characterized by the sudden onset of intense anxiety attacks.

How Your Grandparents' Demons May Be Haunting You

Do you suffer from a strange fear that you can't explain? One in 10 people in the United States has a phobia, which is a crippling fear so intense that it occurs even when there is no danger. For example, people with acrophobia are terrified of heights, even if they are perfectly safe in a high-rise building. In the show *Arrested Development*, Tobias Fünke is "never nude" and even showers with his shorts on; he suffers from nudophobia, which really is a thing. Some other phobias include arachibutyrophobia (the fear of peanut butter sticking to the roof of your mouth), consecotaleophobia (fear of chopsticks), aulophobia (fear of flutes), and my wife's kryptonite, chilopodophobia (fear of centipedes).

According to the National Institutes of Health, more than 19 million people in the United States experience diminished quality of life due to irrational fears and anxiety. Where do these odd phobias originate? Perhaps from your own experience. As a toddler, I nearly choked after shoving too many peanut butter cups into my mouth at once. I'm not sure I qualify as arachibutyrophobic, but to this day I eat peanut butter products with utmost care. Fears and other phobias that can't be explained by our personal experiences may be rooted in our family tree. In *The Graveyard Book*, author Neil Gaiman wrote, "Fear is contagious. You can catch it." And as it turns out, it's not only contagious; it can also spread down through the generations, as suggested in the following experiment.

Acetophenone is a chemical found naturally in many foods, including apricots, apples, and bananas; in its purified form, it smells like cherries. Mice like the smell of acetophenone, but can also be taught to fear the scent. In 2013, neuroscientists Kerry Ressler and Brian Dias at Emory University used mild electrical shocks to condition mice to be afraid of a cherry scent. They would release acetophenone vapors into a cage of mice and then deliver a small shock into their paws through the floor. After three days, a whiff of acetophenone made these mice cower in fear even after they no longer received a shock. In other words, Ressler and Dias instilled an unnatural fear of cherries into these mice.

The researchers then took cherry-phobic males and mated them with normal females that were not trained to fear the smell of cherries. Unexpectedly, their mouse pups were born with a much greater sensitivity to cherry odors, growing anxious and frightful as soon as they caught wind of the scent. This is remarkable, because these pups were never taught to associate cherry odors with a shock. They were simply born that way, as if their father whispered a warning to the pups while still in their mother's womb: "Little dude, run away if you smell cherries or you might get shocked by these goofy people in lab coats."

Passing on a learned fear to your kid is pretty wild. But the biggest "shock" came when the *grandchildren* of these cherry-phobic mice continued to show an enhanced fear response to the cherry scent, even though their parents were never shocked. This surprising passage of traits through multiple generations that doesn't involve changes to gene sequences is called transgenerational epigenetic inheritance.

Why is this so surprising? Because children do not normally inherit things that their parents have learned. For example, I was taught to clean up my crumbs after eating, but I assure you that my kids did not inherit this behavior. I took differential calculus in college for some reason, yet my kids still had to be taught basic arithmetic. And what parent hasn't wished their kids were born potty-trained? So how can a mouse that learned to fear cherries pass this behavior on to his young?

The cherry-fearing pups were born as if they were tipped off to the environment—like a friend texting about a pop quiz occurring later that day. For information like this to be transmitted, some kind of message had to be passed to the sex cells in real time. This, in turn, means that sperm and egg cells must have a means of sensing the environment. The sex cells must also have a means to preprogram their DNA contents, so that the baby is born equipped to survive in their parents' environment. As it happens, sperm cells do contain sensors for hormones, neurotransmitters, growth factors, and, yes, odors. A sperm cell literally "nose" what is going on out there! Could these receptors on sperm be like broad-range scanners that sense potential problems in the environment so the gene crew can prepare before the launch sequence? If so, how does it work?

Ressler and Dias examined whether the fear conditioning changed the mouse's brain in any way. And lo and behold, the cherry-phobic mice were making much more acetophenone odor receptors (named OLFR151) in their brain compared to the unconditioned mice and their pups. The increased number of OLFR151 receptors made the mice more sensitive to the cherry scent.

Ressler and Dias hypothesized that fear conditioning must have caused alterations in sex cells, too. Indeed, they discovered that the OLFR151 odor receptor gene in sperm cells of the cherry-fearing mice had reduced DNA methylation. Without DNA methylation marks in the way, more of this odor receptor is made, leading to a heightened response to the cherry scent in the pups. And when these pups grew up, their sperm also lacked DNA methylation at the OLFR151 gene, explaining why their offspring were still afraid of cherry scents two generations removed from the initial fear conditioning.

Additional examples of transgenerational inheritance have since been discovered in other species, including humans. In previous chapters, we discussed examples of fetal programming, whereby an unborn child's DNA is epigenetically altered in response to a parent's environment, diet, or substance abuse. The same concept applies to transgenerational epigenetic inheritance—except that the changes programmed into DNA are passed on for at least one more generation. One of the best characterized examples of this in humans is seen in descendants of those who survived the Dutch Hunger Winter. In 1944, the German blockade of food to the Netherlands caused a devastating famine, the effects of which can still be seen today. Many of the children and grandchildren of these starved mothers were born small, and grew to be prone to obesity and diabetes. It is believed that mothers who were starved while pregnant had babies whose DNA underwent fetal programming to express genes that maximize extraction of calories from minimal amounts of food, a strategy that makes perfect sense if born into famine. However, thrifty metabolism is maladaptive in times of plenty, which explains why their offspring now struggle with weight problems. More recently, epigenetic marks appear to have been left in the children born of pregnant mothers traumatized by the terrorist attacks of September 11, 2001. The children of mothers who developed post-traumatic stress disorder (PTSD) as a result of the 9/11 attacks displayed an increased risk of anxiety problems.

Both fetal programming and transgenerational epigenetic inheritance are as humbling as they are startling. These corollaries to evolution imply that some of your behaviors (good and bad) could be the result of something your parents, or even your grandparents, experienced. We can put this knowledge to good use: Knowing that stress and trauma may scar DNA for multiple generations should further inspire us to provide better environments for children without delay.

Why Man's Demons Are From Mars, Woman's Demons Are From Venus

In the movie *Ice Age,* two female sloths have a conversation about the lovably geeky character, Sid the Sloth. One laments that Sid is not much to look at, but it's hard to find a family guy. Her friend replies, "Tell me about it. All the sensitive ones get eaten!" This is just one issue that stems from the different evolutionary demands that were placed on each gender in our distant past.

Evolutionary psychologists postulate that all of our behaviors, from the divine to the grotesque, are motivated by the subconscious urge to recruit top-notch mates for reproduction. We go to extraordinary (and often silly) lengths to attract a mate and keep our chain of DNA unbroken, from putting on makeup to putting on a show, from pumping up biceps to a push-up bra. Flaunting our strength, beauty, resources, and number of Instagram followers are requisite in the evolutionary mating game.

Our selfish genes want what they think is best for them, and they seduce us to seek out the most promising mate for a chromosomal merger. That's where the commonalities between the sexes seem to end and some important differences begin, based on inherent disparities in reproductive biology. Evolutionary psychologists assert that, on average, men gravitated toward young virgins because they're at peak fertility and unburdened with a competitor's progeny. In contrast, women had a thing for males of strong stature and rank, as such men have proven capable of acquiring resources that her offspring could use (such men are usually older). Many people still seek out mates based on these Paleolithic criteria. But (lucky me) the modern world doesn't require Popeye arms for reproductive success.

The biological imperative to replicate our genes spawns some of our most desperate and devilish behavior. Evolutionary psychologists theorize that both sexes can be equally vicious to their perceived competition, but in very different ways.

Men and women view sex and relationships differently, largely because of the characteristics of our respective sex cells. Sperm, which are manufactured by the millions and can be replenished quickly, are like Black Friday shoppers, pushing and shoving one another in a ferocious competition to reach a scarce prize: the precious egg. After a successful sale, the sperm-maker is technically free to shop at other stores almost immediately. But the egg-maker goes out of business for at least nine months. A man has the ability to create multiple children in a single day (admittedly, a *really* good day), but a woman is limited to creating one child a year at best. However, a woman has the advantage of knowing that her child definitely carries her genes; a man, not so much. Due to these differences, males take a "quantity" approach to reproduction and females take a "quality" approach. Some argue this is why men are typically more promiscuous than women.

Another consequence of this reproductive economy is that men are generally more physically aggressive and women are more passive-aggressive. A man can afford the risk of physical confrontation, because he's likely already cracked an egg and the mother will care for his young should he lose his life. In contrast, if a woman risks bodily harm or death, this means her young are likely to perish. Women still need to fend off rivals who want the same alpha male, so in lieu of a risky physical fight, nonphysical tactics like rumors, manipulation, and gossip are deployed as weapons. Calling a competitor a "slut" is one of the most damaging rumors a woman can spread; most men avoid a slut's store as it would already be filled with other shoppers, increasing the odds that her children carry another bloke's genes, and not his. Constant concern about the welfare of the limited number of children she can have, along with vigilance required to monitor the rumor mill, creates a great deal of stress for women (and, some argue, has helped women evolve to be more effective multitaskers with superior social skills).

These ideas have led some evolutionary psychologists to summarize that—as a general rule—men are warriors and women are worriers. Not everyone fits this bill, of course. Patty Smyth of Scandal scored a hit proclaiming that she was "The Warrior." (She did not sing, "I am the worrier.")

The differences in reproductive biology also help explain other demons that frequently plague men more than women, from jealousy and stalking to the vile assaults revealed by the #MeToo movement initiated in 2017. Speaking to the magnitude of our reproductive concerns, obsessions with fidelity even besieged the sensitive pacifist John Lennon, who lamented the sour emotions stirring within him in the song, "Jealous Guy." Men evolved suspicious minds because they could never be certain that the children they were fathering were actually their own (that is, until *The Maury Povich Show* came along). Women can also be jealous, but for different reasons. Women are more inclined to forgive their man's sexual transgressions if they were unemotional flings. But if she suspects emotional attachment, then all hell breaks loose. It doesn't hurt her reproductive fitness if the man makes a meaningless booty call, but it could seriously damage her and her children if her man diverts his resources to another woman.

A man can behave in the ugliest of ways in an attempt to ensure his mate remains faithful to him and that her children share his DNA and not that of another man. From this stems possessiveness, bullying, threats, and physical abuse. Because eggs are way scarcer than sperm, men will go to extremes to maximize their chances of having been the winner that fertilized one. If he fails to garner the status and resources necessary to attract a mate, he may in some cases be inclined to cheat at the evolutionary game and resort to the most reprehensible of violations: assault and rape.

Some of this sounds sexist and outdated to modern ears, and indeed there are many exceptions and caveats to the generalizing assertions that evolutionary psychologists propose. Nonetheless, the drives to survive and reproduce are the evolutionary winds that shaped the bedrock of human nature. The biological differences between male and female reproduction still reverberate in our society today, and likely influence some of our stereotypical gender-specific behaviors. It is also possible that these stereotypes unwittingly bias evolutionary psychologists when they devise theories about our evolutionary past.

Are Demons Hiding in Your Genes?

In 1983's "Authority Song," John Mellencamp lamented that he was always losing his fights with the Man. Fast-forward to the present day and you'll see the apple hasn't fallen far from the tree: Mellencamp's two sons have been in trouble with the law on multiple occasions for fighting and resisting arrest. Mellencamp could record a remix of the song, singing, "My sons fight authority, authority still always wins."

The Mellencamp family is not unique in having tempers go up like paper in fire across generations (some would say it's because they're all from the same small town). But science has shown that belligerence can be in the genes. Many studies comparing identical and fraternal twins have demonstrated that a taste for violence is heritable, and the genetic component is as high as 50 percent. Statistics also show that kids who are raised by a loving adoptive family also have high delinquency rates if their biological parents were criminals.

In 1978, a Dutch woman was at her wit's end dealing with several unruly boys in her family. Distraught by their constant mayhem, she sought help from geneticist Han Brunner at University Hospital in Nijmegen in the Netherlands. The affected men in her family suffered from low intelligence and committed abhorrent acts of violence. One raped his sister. Another tried to run over his boss with a car. Two of the boys were fond of setting houses on fire. After years of painstaking research, in 1993 Brunner discovered that the boys in question shared a genetic defect: a mutation in a gene that encoded an enzyme called monoamine oxidase A (MAOA). As subsequent studies by others began to connect this gene variant to other cases of violence, it has since been dubbed "the warrior gene."

About 15 years after its discovery, the gene in question made history again. In 2006, Bradley Waldroup was drinking and reading his Bible while waiting for his estranged wife and kids to arrive for the weekend. When she arrived, dropped off by a friend, they got into an argument that caused Waldroup to snap. He pulled out a gun and shot his wife's friend eight times in front of her and his four children. Then he told his kids to say goodbye to their mother as he chased her with a machete. Waldroup chopped off one of her fingers in the scuffle, but she managed to escape.

During his trial, Waldroup managed an escape of sorts, too. He became the first person to be spared the death penalty, in part because of his genes. Within Waldroup's DNA is the gene variant of MAOA, the same one Brunner noted in the Dutch boys of mayhem. Waldroup's lawyers effectively argued that this genetic predisposition, coupled with his history of child abuse, left him with little control over his murderous actions.

What does MAOA do that might explain its link to violent behavior? The variant gene produces less functional MAOA enzyme, which is needed to break down serotonin, norepinephrine, and dopamine. People with less MAOA would presumably have abnormally high levels of these neurotransmitters, predisposing them to impulsive mood swings and hostility.

Studies performed in mice support the idea: Mice genetically altered to lack MAOA show higher than normal levels of serotonin and norepinephrine, and behave more aggressively. In other studies, MAOA gene variants have also been linked to social phobias and drug abuse. Interestingly, brain imaging studies performed on carriers of MAOA variants show that a region of the brain involved in our fear response (the amygdala) is overactive while an analytical region (cingulate cortex) is repressed; together, this may indicate that the rationalizing part of the brain has trouble calming down its heightened fear response.

In 2014, the MAOA variant turned up yet again in one of the largest genetic studies of criminals. This research, led by psychiatrist Jari Tiihonen at the Karolinska Institutet in Sweden, combed through the genes of nearly 900 Finnish prisoners. As it turned out, the most violent repeat offenders had mutations in both MAOA and another gene called CDH13, which makes an adhesion protein on neurons that may facilitate brain development and function. The team found that individuals harboring gene mutations in both MAOA and CDH13 were 13 times more likely to be a violent criminal. The warrior gene now had an accomplice.

Some researchers have proposed that MAOA helps explain why men are more violent than women. The MAOA gene lies on the X sex chromosome, which means that women (who have two X chromosomes) possess two versions of the MAOA gene. Women thus have a backup that could compensate for a mutant version. But instead of a second X chromosome, men have a Y chromosome, which does not include a second copy of MAOA. And other studies have shown that MAOA has different effects on men and women. A 2013 study by Henian Chen at the University of South Florida showed that although the MAOA variant has been associated with malevolence in men, the MAOA variant is linked to happiness in women.

As the field expands, new genes have been associated with aggression and violent behavior (although they don't have catchy nicknames that a Klingon geneticist might use). The COMT gene encodes an enzyme called catechol-O-methyltransferase; one of its jobs is to break down dopamine, the neurotransmitter associated with our reward response and motivation. Like the MAOA variant, the gene variant of COMT also leads to abnormally high levels of dopamine in our brain, which could compromise rational thinking. Numerous studies (but not all) have linked COMT gene variants that produce less of the enzyme to aggressive behavior. When the COMT gene is knocked out in male mice, they have higher dopamine levels and fight more, supporting the idea that this gene helps keep hostile behavior in check.

Genes involving serotonin signaling have also been implicated in violence. Recall that serotonin is a key mood stabilizer that helps quell impulsive, irrational behavior. Scientists have created what they call the "outlaw mouse" by removing the gene for a type of serotonin receptor called 5-HT1B. A normal mouse is a goner if put into a cage with this ferocious mouse lacking 5-HT1B. In people, serotonin levels are generally lower in those who are quarrelsome, likely making it difficult for them to control emotion in social situations.

On the surface, the discovery of genes that predispose people to violence seems like a major breakthrough. Moreover, most of the genes linked to violence make sense in that they would conceivably alter normal brain chemistry and function. So why aren't we screening people to see if they carry these variants?

Couldn't we pull them out of society before they do harm, as in the Philip K. Dick story "The Minority Report?"

It's not that simple. Some of these gene variants exist in people who wouldn't hurt a fly, and some violent criminals don't possess these variants. As such, many scientists argue that we must stop assigning a misleading behavioral label (for example, "warrior") to single genes. We said it before, but it's worth repeating: Genes encode for proteins, not behaviors. These genetic associations warrant further study, but bear in mind that a gene is one piece of the puzzle that makes up a picture. Just as you can't tell what that picture is going to be solely by looking at one puzzle piece, you can't anticipate someone's behavior based on one gene.

How Your Childhood Demons Affect You in Adulthood

As we've seen with other complex behaviors, genes alone do not accurately predict a person's destiny. The environment is crucial in how that genetic program unfolds, a concept captured perfectly in the *Star Trek: The Next Generation* movie *Nemesis*.

The villain in the movie is a clone of our hero, Captain Jean-Luc Picard. His clone, named Shinzon, was raised in a brutal labor camp, where darkness, loneliness, and torture were the norm. Despite being identical to Picard gene for gene, Shinzon suffered adverse childhood experiences (ACEs) that made him into an ambitious dictator hell-bent on destruction; Picard's well-adjusted upbringing on Earth made him an ambitious explorer and peacemaker.

It is profoundly humbling to realize that if we were raised in different circumstances, we would not be the people we know today. And because we had no control over the genes or the childhood environment we were dealt, both of which influenced how our brain turned out, then just how accountable are we for our behavior?

To this point, as many as 30 percent of people express the variant form of MAOA, and yet most of them do not turn into Hannibal Lecter. Some studies have shown that when someone with the MAOA variant is also subjected to ACEs (particularly child abuse, as in the case of Bradley Waldroup), they become significantly predisposed toward impulsively violent behavior. In the Finnish prisoner study mentioned previously, the researchers failed to find a greater proclivity for violence in those with the MAOA variant who were also mistreated as children. But they did find that alcohol or amphetamine use greatly increased impulsive aggression among carriers. So although it is evident that environment plays a substantial role in whether the MAOA variant translates to increased aggression or violent behavior, more research is needed to figure out exactly how.

As we've seen in previous chapters, the dawning science of epigenetics is showing that ACEs do more than just psychological damage. They also chemically alter the structure of DNA and how genes are expressed. Even ACEs that some people still consider to be a minor and a normal part of growing up, such as bullying, leave an alarming mark on victims' DNA. Psychologist Isabelle Ouellet-Morin at the University of Montreal has shown that kids who are bullied become desensitized to stress and have a higher likelihood of growing up to be socially inept and aggressive themselves. In a 2013 study, her team showed that the blunted stress response in children who were bullied is associated with increased DNA methylation at their serotonin transporter gene, shutting it down. As we've seen before, serotonin regulates mood and is involved in depression. Sticks and stones can break your bones, but bullying can also break your DNA.

It has long been suspected that poor nutrition in utero or during childhood contributes to persistent behavioral problems in adulthood. Data collected from the Dutch Hunger Winter famine of World War II confirmed that men exposed prenatally to severe maternal nutritional deficiency during the first and/or second trimesters of pregnancy exhibited increased risk of antisocial personality disorder. Poor nutrition during pregnancy signals to a growing fetus that he or she is about to be born into a stressful, resource-deprived environment; consequently, fetal programming sets appetite genes to be metabolically thrifty and stress response genes to be in a heightened state of alert. Such traits may be useful in stressful environments, but can be maladaptive if the environment improves.

In the United States, we are faced with the opposite problem: obesity. Despite the abundance of food, many are still starving for essential vitamins and minerals that diets high in sugar, fat, and salt fail to provide, and these deficits can lead to conduct disorders. For example, zinc and iron deficiencies are found in many juvenile offenders. Young prisoners at the Aylesbury jail in the U.K. who were given vitamin and mineral supplements committed 37 percent fewer violent offenses within the prison.

Low omega-3 fatty acid levels have also been linked to aggression. It's not as fishy as it sounds: Omega-3 fatty acids play important roles in the function of the brain, and a 2015 trial by neurocriminologist Adrian Raine at the University of Pennsylvania showed a reduction in behavior problems with omega-3 supplementation in children aged 8 to 16 years. Studies by other researchers have found that countries with low homicide rates, like Japan, eat many more kettles of fish (a rich source of omega-3). A 2007 study showed that women who ate more than 340 grams of fish a week while pregnant had children with better social development and IQ scores. The writing is on the wall: Proper nutrition must be sustained throughout childhood and adolescence to ensure proper brain development.

Exposure to toxins at a young age also affects gene expression and brain development in ways that could produce a spectrum of behavioral problems. Childhood exposure to such environmental toxins as lead may be a grossly underappreciated contributor to America's problem with violent crime.

Getting lead poisoning can be easy, because lead can be inhaled, absorbed, or ingested; even small amounts can cause irreversible damage. In the body, lead can slip into the cockpit of proteins, sitting where minerals like calcium, iron, and zinc are supposed to be. This has disastrous consequences on several bodily systems, including the brain, where calcium is used to transmit electrical impulses. Thus, lead can cause mental problems such as impulsivity, attention disorder, and learning disabilities, setting a stage for antisocial and violent behavior in adulthood. And the proof is in the pudding: Heavy metal poisoning is suspected to be a major contributing factor in numerous incidents of insane behavior through the ages, from Van Gogh slicing off his ear in 1888 to the 1984 San Ysidro McDonald's massacre (in which the shooter was a welder who had lead poisoning and the highest levels of cadmium ever recorded).

Elegant studies that compared crime statistics between cities that used lead pipes and those that did not show a striking correlation between childhood lead exposure and violent crime. Another study examined kids who lived in the same city but were exposed to different levels of lead. Here, researchers compared kids who lived closer to roads before lead was removed from gasoline with kids who grew up away from such roads or during the time when lead was removed from gasoline. The children exposed to the higher levels of lead had higher rates of school detention and suspension.

In addition to the well-characterized effects of lead short-circuiting brain signaling, researchers also found that early exposure to lead alters DNA methylation patterns in genes linked to developmental and neurological disorders. The effects of heavy metal poisoning may be felt for generations, as suggested in a 2015 study that showed DNA methylation changes in the grandchildren of mothers who were exposed to lead.

Most people think that lead poisoning is an old problem from the 1970s, when kids in bell-bottoms munched on lead paint chips. However, the liberal use of lead in buildings, gasoline, and water supply pipes decades ago still haunts us today. In 2014, an inexcusable calamity in Flint, Michigan, reminded everyone of the effects of acute lead poisoning. After officials switched the city's water supply from Lake Huron to the Flint River in a cost-cutting measure, the 100,000 citizens of Flint unwittingly consumed large doses of lead that produced serious health problems. Given that lead persists in the body for years and may even have transgenerational effects, it is feared that additional cognitive and behavioral consequences may be seen in upcoming decades.

A similar problem may have already occurred in Chicago, which at the time of this writing is struggling with an unprecedented epidemic of violence. Some scientists believe that the explosion of violence in Chicago today may be caused, in part, by lead poisoning that occurred in 1995, when over 80 percent of the children in today's most crime-ridden neighborhoods tested positive for dangerously high levels of lead. Perhaps the best way to get tough on crime is to get tough on those who commit crimes against our environment.

Alcohol and other drugs are additional poisons that can instill demons in one's brain before their first breath is taken. In the United States, where up to a quarter of pregnant women still smoke, boys have four times the risk of being a problem child if their mother puffed 10 cigarettes a day while pregnant, and girls have five times the risk of drug addiction. Even pregnant women exposed to secondhand smoke have a higher risk of their children growing up to have conduct disorders.

Smoking causes higher than normal testosterone levels during pregnancy, which can prime a fetus for a life of misbehavior. A curious effect of high prenatal testosterone is that it causes the ring finger to be longer then the index finger. It doesn't apply in all cases, but many studies have correlated longer ring fingers to higher levels of dominance, impulsivity, and aggression. Smoking may also induce some of these adverse behavioral changes through fetal gene programming, as tobacco can alter DNA methylation in utero. It has also been shown that nicotine interferes with blood flow in the uterus, which slows the rate of oxygen getting to the fetus, putting it at risk for brain damage.

If a mother boozes during pregnancy, her child can suffer from fetal alcohol syndrome (FAS). Roughly one child is born with FAS for every 1,000 births. FAS can cause a number of physical and mental impairments, particularly with respect to how to manage social interactions. Adolescents and adults with FAS are unresponsive to social cues, fail to reciprocate friendships, lack tact, and have trouble cooperating with people.

We tend to write off such people as jerks, but their boorish behavior may not be their fault. Given their problems recognizing social norms, it is not surprising that more than half the people with FAS have trouble with the law. Drinking even small amounts of alcohol while pregnant can triple the chances of delinquent behavior in the unborn child. Sometimes, a child can be born with symptoms resembling FAS even if the mother never had a drop of alcohol. How can that be? The blame goes to the father. Heavy drinking can alter DNA methylation patterns in the sperm of the father-to-be at genes that are important for prenatal development.

Collectively, these studies suggest that some criminals may have been childhood victims of poisoning by an agent that disrupted normal brain function. These toxic agents can be psychological in the form of parental abuse or bullying by peers, or physiological in the form of poisons like heavy metals, nicotine, or alcohol. In any case, it's clear these insults can sow the seeds of antisocial behavior, aggression, and violence by directly interfering with brain development and signaling, or through epigenetic programming of the individual's DNA.

How Brain Invaders Drive You Mad

You might feel rage welling up from the pit of your stomach. But in reality, such feelings are actually meted out by the brain. In a dramatic 1963 experiment that truly makes living things appear to be no more than meat robots, Yale University physiologist José Manuel Rodriguez Delgado used a remote control to stop a bull that was charging right for him. Before the demonstration, Delgado implanted a small device in the bull's brain that would emit electrical impulses when he pressed the remote control; the electrical impulses mimic what normally happens when neurons communicate with one another. By stimulating a specific part of the brain with the touch of a button, Delgado had literally stopped the bull's aggressive instincts in their tracks.

Humans are not immune to this sort of brain control; depending on which part of the brain is electrically stimulated, people can experience any emotional state, including outbursts of laughter, tears, or anger. Scientists like Mary Boggiano at the University of Alabama at Birmingham are using variations on Delgado's technique to quell impulsive behavior, such as binge eating. When someone gets a craving to overeat, an electrical current is sent into the brain to stop the urge in its tracks, just like Delgado's bull.

The brain is a delicate piece of machinery and, although encased in a fairly sturdy container, can be offset by a wide array of insults. The infamous clock tower shootings that took place at the University of Texas in Austin in 1966, which killed 16 people and wounded 31 others, were likely caused by a clump of cells no bigger than a pecan.

Charles Whitman was the prototypical all-American kid and Eagle Scout. But when he turned 25, he began to get tremendous headaches. He went to see campus health services because he was having disturbing thoughts that he could not control. Further clues of his mental deterioration were left behind in notes he wrote before his killing spree; Whitman was so convinced he was losing his mind that he requested autopsy doctors to examine his brain and donated his money to mental health research. Sure enough, the autopsy revealed a tumor pressing up against his amygdala, the brain region critical for regulating fear and anxiety.

Additional types of brain injury have been associated with good people gone bad, including damage caused by stroke, concussion, or infection. Children or spouses who are abused often sustain brain injuries that lead to aggressive behavior. Contact sports like football frequently produce dangerous concussions that have been linked to triggering bouts of violence; the term "punch-drunk" originated with boxers who exhibited cognitive defects from years of repeated blows to the head. Punch-drunk is now known to be a subtype of chronic traumatic encephalopathy (CTE), a worrying condition that appears to affect more athletes playing contact sports than previously thought.

CTE was first linked to football players by Dr. Bennet Omalu, whose story was featured in the Will Smith film *Concussion*. It correlates with a loss of impulse control, erratic behavior, and aggression, most likely explaining why some of its sufferers undergo a tragic transformation into a monster. High-profile cases include Kansas City Chiefs linebacker Jovan Belcher, who killed his girlfriend before taking his own life in 2012. Aaron Hernandez, who played for the New England Patriots, was diagnosed postmortem with the worst case of CTE ever seen in someone under 30. Hernandez was serving a life sentence for a 2013 murder but hanged himself in prison in 2017. Ties between CTE and violence are certainly not limited to football; in 2007, professional wrestler Chris Benoit murdered his wife and seven-year-old son before hanging himself from his weight lifting machine. We should take a good, hard look at whether to continue letting our children damage their brains by head-butting soccer balls or playing tackle football.

In addition to tumors and tissue damage, there is another type of insidious brain invader: microbes. When confronting our demons, we usually don't give much thought to these little devils. Perhaps the most well-known pathogen that causes aggression is the rabies virus, whose name derives from a phrase meaning "to do violence." The viral particles that cause rabies are introduced into a new victim by surfing a wave of saliva. Rabies commandeers the brain and turns the infected animal into a ferocious beast with an insatiable appetite for flesh. By manipulating its victim into biting others, rabies can spread to other animals.

Although rabies announces its presence in the brain with all the subtlety of Lady Gaga, the single-celled parasite *Toxoplasma gondii* prefers to lay low. Recall that *Toxoplasma* stealthily makes its way to the brain of any warm-blooded animal it infects (including the three billion people carrying this parasite) and sits there for the rest of the host's life in the form of latent tissue cysts. As disquieting as that thought may be, it has long been believed that these cysts are benign, causing a problem only in people with weakened immune systems. But that assumption was shattered in the 1990s, when Joanne Webster, then at the University of Oxford, noted strange things happening in rats she infected with *Toxoplasma*. Remarkably, the *Toxoplasma*-ridden rats lost their innate fear of cat odors; in fact, the infected rats seemed *drawn* to the scent of their predator. Webster dubbed this phenomenon "fatal feline attraction."

From an evolutionary perspective, this makes perfect sense, as felines are the only organism that supports the sexual stage of the parasite. Only when *Toxoplasma* finds itself in the romantic milieu of cat entrails does it turn on the Marvin Gaye and get it on. In other words, this parasite does something to the rodent brain that transforms the creature into a taxi that takes them to the love shack. The infected cat will then excrete billions of infectious *Toxoplasma* oocysts into the litterbox, sandbox, garden, and streams. These oocysts have thoroughly contaminated the food and water chain, explaining why so many people have *Toxoplasma* in their brain.

If *Toxoplasma* can manipulate a rodent brain, how might it affect our own? Some have speculated that if *Toxoplasma* makes rodents attracted to cats (perhaps this parasitic infection explains the "crazy cat lady" phenomenon). As mentioned in Chapter 4, correlative studies suggest some general trends seen in people infected with the parasite compared with those who are not. One of the strongest correlations is the link between *Toxoplasma* infection and the development of neurological anomalies, especially schizophrenia. People carrying *Toxoplasma* tend to be more anxious and open to risk taking, but some gender-specific differences have also been documented. Infected men tend to be more introverted, suspicious, and rebellious, and infected women tend to be more extraverted, trusting, and obedient.

Could *Toxoplasma* be yet another factor that awakens our dark side? In a 2016 study, behavioral neuroscientist Emil Coccaro at the University of Chicago found that people infected with *Toxoplasma* are twice as likely to have intermittent explosive disorder, a condition whereby sufferers are liable to irrational outbursts of aggression with little provocation.

Is There a Devil Inside Us?

Susannah Cahalan led a normal life until she turned 24 in 2009, when bizarre problems began to plague her. From out of nowhere, she began having problems speaking; her tongue became easily twisted into knots. Next, Cahalan experienced problems with mobility, lurching around like the bride of Frankenstein. In addition to these physical problems, she became paranoid and violent. She suffered hallucinations and adopted other personalities. She was convinced that her father had murdered her stepmother. Cahalan descended rapidly into madness, uttering otherworldly noises and reaching a catatonic state within a month.

The sudden transformation of this young, vibrant woman was truly terrifying and defied all logic. She had no head injury, brain tumor, infection, or toxin in her system; no medication for mental illness had helped her. With known culprits ruled out, what other possible explanation could there be besides demonic possession?

Fortunately, her family called a neurologist instead of an exorcist. With one simple test, Souhel Najjar was able to diagnose Cahalan's condition. He asked her to draw a clock. Intriguingly, Cahalan put all the numbers on just one side of the clock's face, which indicated that her brain was malfunctioning. Najjar suspected inflammation and described the illness as "brain on fire," a phrase that became the title of Cahalan's memoir of the maddening experience. Much to the chagrin of the International Association of Exorcists, Cahalan's condition was not caused by an evil spirit; rather, it had a purely biological explanation, just like any other strange neurological anomaly. If Cahalan had not been diagnosed, she most likely would have suffered irreversible brain damage or even slipped into a coma and died.

The disease afflicting Cahalan was first documented just two years before her case, and it is called anti-NMDA receptor encephalitis. As early as 2005, neurologist Josep Dalmau had been studying a group of patients exhibiting the same haunting symptoms that gripped Cahalan. To gain insight into what might be going on, he took samples of their blood and cerebral spinal fluid and put them onto sections of rat tissues. He discovered that these "possessed" patients possessed a substance that stuck to the brain, specifically to proteins called NMDA receptors found on the surface of neurons.

NMDA stands for N-methyl-D-aspartate, which is a chemical we produce that acts in the brain. The NMDA receptor is important for memory and learning, and it helps nerve cells communicate with one another. For reasons that have yet to be fully resolved, some unfortunate people start making antibodies to this receptor. Our immune system normally makes antibodies to fight off foreign invaders. But sometimes the body makes antibodies against a piece of us (hence the name "autoimmune" disease). It is like nonstop friendly fire in your own body, and it can take a devastating toll. Cahalan's autoimmune disease represented a new type of brain injury that masquerades as a demonic force.

Neurotransmitters critical for signaling between brain cells work through NMDA receptors but can't bind to them if antibodies are blocking their access. By disrupting neuronal signaling, the anti-NMDA receptor antibodies were creating chaos in the brain that led to Cahalan's psychiatric symptoms. Since the discovery of this disease, many others with varying degrees of psychoses have been diagnosed. Case reports include paranoia, hallucinations, bodily harm to oneself and others, obsessive thoughts, uncontrolled movements, speaking in tongues, seizures, catatonic states, and other eerie behaviors. Not all patients with this disease have seen a happy ending, but Cahalan made a full recovery after receiving immunosuppressive treatments. These medications dampen the immune response, dousing the friendly fire by taking away the ammunition. By shutting off her body's ability to make anti-NMDA receptor antibodies, neuronal signaling was restored to normal. Cahalan's case teaches us that science is the elixir that allows us to rise above our demons.

Should We Feel Sympathy for the Devil?

Like Rudolph's nose, science is lighting our way through the fog, dispelling even the darkest mystiques of our psyche. We no longer have to settle for meaningless and unhelpful explanations like evil or possessed souls. Our fears and demons arise from a cauldron of factors that include genetic predispositions, fetal programming, our evolutionary heritage, and transgenerational epigenetic inheritance. People who turn to the dark side are not consumed by mean spirits, but may have been co-opted by malnutrition, heavy metal poisoning, a head injury, infection, or autoimmune disease. The take-home message is that our demons are not otherworldly; they are wholly rooted in biology. As we begin to unravel the biological reasons why people engage in wrongdoing, we will find more effective means to prevent crime and rehabilitate offenders. The only true sin left to commit is ignoring these facts.

Bradley Waldroup murdered a woman, savagely attacked his wife, and traumatized his children. Just typing that makes my fingers curl into a fist. My primal instincts want revenge on the order of *Django Unchained*. But as we've seen in earlier chapters, our instincts are often wrong and must be reevaluated with reason and objectivity. It is natural and appropriate to feel sorrow for the innocent victims of crime. But does Waldroup deserve some sympathy too? Is it possible to pity a murderer like him without subtracting from the sadness we feel for the victim? Do we have enough tears to weep for both?

This is not to say that violent offenders are entitled to a Get Out of Jail Free card. But if you care about solving the problem of violence, you must care about the offenders. Think about all the things that happened to Waldroup that were beyond his control. He was a victim of ghastly child abuse, which we know is a major risk factor of future behavior problems (in part because of epigenetic changes dysregulating the stress response). He also suffered from depression and rage disorder, conditions that could be the result of genes, microbiota, parasitic infection, or a combination thereof. He may have been genetically predisposed to aggression, further exacerbated by the adverse childhood events to which he was subjected. And he may also have been genetically predisposed to alcoholism, yet another factor in his actions on that fateful night.

Waldroup was caught in an unfortunate storm, a storm that would have drowned most anyone in the same boat. There is little hope if we lazily dismiss criminals as evil souls. But if we can feel even a glimmer of empathy for Waldroup, we've taken the first step onto a more productive avenue to prevent future tragedies.

Until society develops effective means to ensure that every child is raised in a safe and nurturing environment, we are simply inviting future criminal activity. Whether we are talking about petty thieves, murderers, or terrorists, we need to ask: Do we want to wait and punish them as adults or are we willing to help them while they're children?

» CHAPTER SEVEN «

MEET YOUR MATCH

I gave her my heart, she gave me a pen.

—Lloyd Dobler, *Say Anything*

It was 1980-something and my teenage body was undergoing a metamorphosis. I'd seen naked women before puberty (mostly in the *National Geographic* magazines at my elementary school library), but they weren't much more than a funny curiosity. By the time I saw the movie *Porky's* on cable TV, I would experience a new sensation. Until that moment, I figured my private parts had just one job: to get rid of all that soda I drank. But the pubescent hormones surging through my veins turned them into an onboard entertainment system. Like a force awakened, I suddenly felt intense attraction toward the opposite sex. It was not a feeling I could control, nor one that I consciously chose.

I needed help navigating this baffling new world of teenage romance, so I went to my favorite teacher: music. How elated I was to learn that so many famous singers were just as confused about love! Howard Jones inquired "What Is Love?" Van Halen questioned "Why Can't This Be Love?" Tina Turner asked, "What's Love Got to Do With It?" And both Survivor and Whitesnake wondered "Is This Love?" Hearing songs like "Love Is a Battlefield," "Maneater," and "You Give Love a Bad Name" filled me with great anxiety about approaching girls. I didn't want to go to war, get chewed up, and be shot through the heart.

My records taught me that love is majestic and magical, but my biology teacher blinded me with science. In class, we learned that love is really just a covert operation orchestrated by selfish genes that trick us into protecting their legacy: not quite what you read in a Valentine's Day card. I was raised believing that love was a matter of the heart, but I've come to learn that it's all in our head. Love is 50 shades of gray matter. Organisms that don't have a brain, such as bacteria, sea squirts, and many politicians, thrive without this crazy little thing called love. So why does reproduction have to be so complicated for us?

Why It Takes Two, Baby

Bacteria and amoebas have it easy. To reproduce, they simply clone themselves. They don't need to scroll through profiles of potential mates, wondering how much liberty the person took with the truth. They don't need to dress to the nines, marinate in perfume, and feign interest in someone's mind-numbingly dull hobby while nibbling at an overpriced candlelight dinner. The only thing bacteria need to unzip is their DNA; as it peels apart, enzymes make a copy of it to stuff into the daughter bacterium as she buds off the parent. No cuddling, no messy sheets, no confessing that the only thing you know how to make for breakfast is Pop-Tarts.

Not only is bacterial replication easy; it is way more productive. One bacterium can split into two in about 30 minutes, then those two into four, then those four into eight, and so on. Bacteria can have millions of babies by the morning and never have to ask, "Was it good for you?" So why did nature even bother with inventing sex?

The key benefit of sex from an evolutionary perspective is genetic diversity. Asexual replication produces clones. Aside from a random mutation every now and then due to a DNA copying error, the daughter bacterium will be identical to the mother bacterium. Through the lens of selfish genes, that is the ultimate replication strategy. But there's a catch: If the bacteria encounter a threat—say, a mold that secretes penicillin—their entire clonal colony could be vanquished. The clonal colony next door, however, may be resistant to penicillin because they have a gene that makes an enzyme that destroys the antibiotic. If only there was a way to get that gene! That's where sex comes in. Bacteria can partake in a form of sex called conjugation, whereby one bacterium gives DNA to another through a tube called a pilus, which it erects and inserts into another bacterium. Sound familiar?

Sex originated to exchange genes like trading cards, which amounts to a considerable compromise for selfish genes. Instead of 100 percent of the genes being passed into the next generation, only 50 percent get passed. The other 50 percent come from the sexual partner. Sex dilutes an individual's genes, but the resulting mix creates variation in the survival machine a new combination of DNA will build.

Why is variation important? One of the leading ideas is called the Red Queen hypothesis, which draws its name from the children's classic *Through the Looking-Glass*. But instead of Alice racing the Red Queen, scientists believe that organisms race against the parasites that infect them. Think of your body as one survival machine and parasites as another. We are constantly locked in an evolutionary arms race against germs. If we become resistant to a germ, it usually adapts before long and becomes a threat once again. To help ward off infection, it does a species good to keep shuffling the genes in its deck.

Evidence for the Red Queen hypothesis comes from setting up cage match competitions between organisms and their parasites. While at Indiana University, biologist Levi Morran put roundworms called *Caenorhabditis elegans* into the ring with a bacterial pathogen called *Serratia marcescens* and watched them duke it out. Roundworms can reproduce with or without sex, and experimenters can control whether they get physical or go at it alone. The roundworms forced to reproduce without sex lost the fight to the bacteria in just 20 generations. However, roundworms that were allowed to sexually reproduce did not succumb to the bacterial infection. The next time you have sex, be sure to take a moment and pay homage to the germs that made it possible.

Why We're So Superficial

Sex is an advantageous compromise for selfish genes, but a compromise nonetheless. Selfish genes need intel to identify the best partner for a genetic merger, so they tweaked certain physical traits into DNA billboards. Much like the commercials competing car salesmen make, some of these DNA advertisements have evolved to be loud and obnoxious.

Physical features are the first clues we get when sizing up a mate, providing a crude but fast readout for the relative quality of that individual's genes. Across the animal kingdom, species of all kinds use these physical cues to audition a potential mate, and they often dictate whether you swipe left or right. Selective pressure has driven some of these features to absurdity. The most familiar example is the beautiful, yet spectacularly impractical tail feathers of the male peacock. Such extravagant traits puzzled even Charles Darwin, as they seemed like a waste of energy that burdened the bird and exposed it to predators.

Darwin solved this riddle with the idea of sexual selection, which posited that organisms sport seemingly useless features purely to increase their attractiveness to the opposite sex. Females would perceive a male peacock with a garish fan of feathers as a good catch. If he can support such a ponderous plume, yet still evade predators, then he must be exceptionally strong and cunning—characteristics that females may perceive as beneficial for her offspring. Alternatively, the dazzling tail display serves as a glaring signal that the male is ready for mating; the more brilliant the display, the more likely he (and his male offspring) will attract mates. If she is to dilute her genes, this peacock's DNA is worthy to swim in her gene pool.

People exhibit certain features that may be subject to sexual selection as well, such as facial and bodily symmetry. There's a reason why we initially recoil at characters like Sloth from *The Goonies:* We subconsciously associate asymmetry with adverse health. These preferences appear to be preset at birth, as infants just a few months old would rather gaze at symmetrical and attractive faces. Despite our righteous campaigns that teach not to judge a book by its cover, being attractive still has a lot of perks. Studies show that more symmetrical men mate earlier in life, get more girls, and even tend to give their female partners more frequent orgasms. Plus, it would seem that size does matter, though not in the way you might think; men with bigger wallets also tend to attract more mates.

People generally want to tango with those who are physically fit, with smooth and clear skin, healthy white teeth, sparkling eyes, and silky lice-free hair, because the opposite implies the person has bad genes or an infection. Similarly, most people seek out a mate who is energetic, good-natured, intelligent, and cheerful, as these are indicators of good mental health. As the glory days pass us by, our wrinkles and hair loss serve as signals to younger folk that we are no longer spring chickens. Our innate drive to keep up appearances that advertise youthful vitality is so intense that it has birthed multibillion-dollar cosmetic and plastic surgery industries.

Evolutionary psychologists have additional theories about why men and women have historically sought out different qualities in one another. It's no secret that many men use T and A as a proxy for DNA. Men in cultures all over the globe have an uncanny ability to accurately assess a woman's waist-to-hip ratio, the most favored being a waist circumference of 70 percent of the hips. That waist-to-hip ratio is precisely what is ideal for maximum fertility. Studies show that women with figures deviating from these proportions have a harder time getting pregnant, have more miscarriages, and are even more prone to chronic diseases and mental disorders.

Scientists have speculated that men look upon large bosoms like a treasure chest because their primal, subconscious mind equates them with good health and vitality: important qualities in someone who will be nourishing your spawn. In one of the more titillating studies performed to support this idea, men with either an empty or full belly were asked to judge the attractiveness of a bunch of breasts. Results showed that men who were hungry rated larger breasts as significantly more attractive, whereas men who were not hungry did not display this bias. Men also tend to prefer young women because they are more likely to be fertile virgins who aren't spending energy on some other dude's little ones. Women are onto this, and use higher-pitched voices while flirting to sound younger.

In contrast, women are generally more interested in hooking up with a man who has status and wealth, as such resources will be beneficial to her and her young. Although every girl is crazy about a sharp-dressed man, women also fawn over studly looks, preferring a broad-shouldered man with a rugged jaw and well-defined browridges. Masculine features such as these are formed during puberty by high levels of testosterone and provide a woman with a quick and easy readout for strength and power.

Because our male ancestors also required ambition, smarts, and networking skills to gain rank in the social hierarchy, women actively seek these intellectual qualities in their men, too. However, these qualities take longer to assess than chiseled jaws and broad shoulders. And because women have fewer chances to reproduce than men do, this has been proposed as one reason why women generally take longer to decide whether a man is good enough for her.

This kind of evolutionary imperative often translates to cultural attitudes that, for better or worse, remain entrenched. As the old saying goes, women are viewed as sex objects and men as success objects. In 1980s songwriting parlance, men are nasty boys and women are material girls. Despite all social progress to the contrary, many people still behave this way, with these stereotypical patterns revealing themselves soon after puberty as young men and women begin to shop around. In general, teenage boys yearn for the busty cheerleaders and teenage girls get dreamy eyes for jocks with fancy cars. (I can tell you from personal experience that most teenage girls are far less impressed with a neo-maxi zoom dweebie who has collected every single Star Wars Topps trading card and conquered Zork on his Commodore 64.)

People who are superficial in their choice of mates may be so inclined because of all this old evolutionary baggage. But as scientists expose these secrets about our human nature, it is hoped that we will become more aware of the flaws in how our unconscious brain sizes up a mate. Although these simplistic approaches may have served our ancestors well in the past, we have the smarts to rise above the desires of our selfish genes and include inner beauty in our mate selection process. As profoundly unfit as I am to have survived the Paleolithic period, I was still able to find love and affection in this day and age. And, as grotesque and asymmetric as he was, Sloth became a beloved friend and hero to the gang in *The Goonies*.

Why Love Stinks

No one with a crush likes to hear that dreadful consolation prize, "I only like you as a friend." Science is here to tell you that you're taking it all too hard. When someone turns you down, it is probably for a biological reason outside of your control. So don't rush to change your clothes, your hair, or your face; the answer might lie in how you smell.

The types of scents that are important in animal magnetism are pheromones, chemical substances the body releases into the environment to be sensed by other animals. Most animals have a specialized sensor pad in their nose called the vomeronasal organ that relays pheromone messages directly to the brain. Evidence that pheromones operate in humans was first presented in a 1998 study by psychologist Martha McClintock at the University of Chicago, who famously showed that menstrual cycles of women living together become synchronized, thanks to armpit pheromones. Pheromones are a bit creepy in that their activity occurs below our conscious radar. Although men and women engage in awkward small talk to get to know each another, a bevy of chemical information is being sent up our nose and activating subconscious areas of our brain. Ever been chatting it up with a potential partner who seems perfect in every logical way—but then you get a strange feeling that this person is not the one? It wasn't anything they said or did; your brain just says, "I have a bad feeling about this." Maybe you've been on the receiving end of such an exchange. It's not fun in either case. But perhaps you can rest a little better knowing that no one is to blame; it may be pheromones.

The theory that chemicals emitted by our body subconsciously affect our romantic inclinations has been put to the test in numerous ways. Biologist Claus Wedekind at Bern University in Switzerland conducted a classic study in 1995 that involved smelling dirty T-shirts, and revealed that women can sniff out men who possess immunity genes different from their own. The men in this experiment wore cotton T-shirts for two days before the brave women in the study took a good whiff of the armpits and ranked the scent. The results showed that women preferred the smell of T-shirts worn by men who possessed different immune system genes. If she had similar immune system genes, she found the man's scent to be less attractive.

Why is it advantageous to pair up with someone who has different immune system genes? It goes back to the Red Queen hypothesis and why we have sex in the first place. Because our immune system needs to respond to a large number of germs that can mutate quickly, it is beneficial to have a diverse arsenal of immunity genes to tackle these varied germs. There is also evidence that having immune genes that are too similar leads to a higher risk of miscarriage. So being turned down by someone is not personal; it is more like organ rejection.

The scent of a woman matters, too. A woman's ability to attract a man might be augmented when she is in estrus and looking for a new love. If you've ever seen monkeys at the zoo, it's pretty obvious which females are in heat. But it's not so easy to spot human females at the height of their fertility. However, some studies suggest that a woman's body odor fluctuates during her menstrual cycle in ways that men can notice. In 2006 anthropologist Jan Havlíček at Charles University in Prague had women volunteers wear cotton pads in their armpits during different stages of their menstrual cycle. A group of men then smelled the pads and rated the pleasantness of the scent. The result? Odor pads collected from women in their fertile phase were rated as the most alluring. If true, biology has a way of sweetening the pot when our gametes are ready to play.

As if these stealthy odors we emit weren't creepy enough, there is now evidence that diet can affect them, too. You are what you eat, and you attract others who eat what you eat. You're probably thinking that's obvious; a strict vegan is not likely to get along with a carnivore. But what we're talking about here is how diet may affect pheromones through a third party—your microbiota.

A 2010 study by microbiologist Gil Sharon at Tel Aviv University found that the intestinal bacteria in a fruit fly called *Drosophila* are a critical factor in mate selection. Flies that eat a diet of molasses like to fly into the danger zone with other molasses-fed flies, whereas flies that eat a diet of starch prefer to hook up with other starch-fed flies. But if you give the flies antibiotics, which deplete the intestinal bacteria, then anything goes—molasses flies would do the wild thing with starch flies and vice versa. In a style reminiscent of "There Was an Old Lady Who Swallowed a Fly," Sharon and his team discovered that diet affected the intestinal bacteria, which affected the pheromones the flies produce, which affected mate selection. In humans, studies have shown that women prefer the smell of men who eat more vegetables. Being a supertaster, this study explains my lackluster track record with women.

Finally, odors that accompany experiences in our youth can have an eerie influence on us when it comes time to play the dating game. This was first demonstrated in a classic study from 1986. Researchers had newborn male rat pups suckle from a mother who was perfumed with a citrus scent. Upon weaning, they stopped applying the citrus scent to the mother. One hundred days later, they compared how those male rats interacted with unperfumed females or those doused in the citrus scent. In a result that would make Sigmund Freud proud, the citrus-scented females aroused males much more easily if the male's mother smelled of citrus during nursing.

A similar study performed by another group in 2011 allowed juvenile female rats to play with other rats that smelled either of almonds or lemons. When they reached mating age, the females showed a bias toward males that smelled like her juvenile playmates.

Together, these studies suggest that aromatic experiences during infancy and youth can covertly influence what sort of partner sets off the fireworks. If this holds true in humans, fellows who are seeking my daughter's heart stand a better chance if they smell of mac and cheese.

As science is revealing just how important smell is during mate selection, we nevertheless go to great lengths to thwart our natural odor. Many of us shave off the tufts of hair that provide habitats to skin microbes and help broadcast our scent. After scrubbing away our skin microbiota with daily showers, we bathe ourselves in colognes, perfumes, and deodorants. These agents disguise the microbial signals that our body subconsciously uses to evaluate mating candidates. Glossing over this crucial information is like hiring someone without interviewing them first. As someone who routinely drives Boy Scouts home from their weekend camping trips, I'm not advocating that we forfeit the olfactory pleasantries offered by soaps and deodorants. But when it comes to determining if your date is right for you, maybe at least put them through the T-shirt test or stick your head in their hamper when they're not looking.

Why Opposites Attract but Don't Usually Last

Do opposites attract? They most certainly can, but they repel pretty quickly after you cool the engines. We've seen this happen with Sam and Diane from *Cheers,* Han and Leia from *Star Wars,* and Paula Abdul and MC Skat Kat. A 2003 study by behavioral ecologists Peter Buston and Stephen Emlen at Cornell University showed that most people follow a "likes-attract" rule when screening potential mates. The likes-attract rule makes sense in the context of the selfish genes model. If selfish genes must yield half their territory to sexually reproduce, then why not recruit genes similar to those surrendered? Couples stand a greater chance of lasting if they complement each other, like two verses in the same song.

There is a strong tendency for couples to be of similar age, height, body shape, and personality. Anthropologists have a fancy phrase for this called "positive assortative mating," and other animals follow the same principle. Next time your partner asks why you fell in love, look blissfully into their eyes and whisper in the sexiest voice you can muster, "Positive assortative mating, babe."

Positive assortative mating seems to contradict the odor experiments that suggest we unconsciously seek out mates that would diversify our offspring's genome. No one said love was easy! These competing principles act like opposing weights to balance a scale: Your ideal mate should be similar to you, but not too similar. When we slide toward the too similar end of the spectrum, we defeat the purpose of sex, which is to sprinkle variety into the genetic repertoire. This is why we have strong instincts that discourage romantic feelings toward close family members. Avoidance of incest is one of the most universal taboos in human cultures, and it is observed throughout the plant and animal kingdom. It may even explain why brothers and sisters despise one another during their peak fertility window in adolescence.

There's an important biological reason why we are wired to be repulsed by incest: Too much genetic similarity yields offspring that are enriched for deleterious traits. The bad boys of the genome are not weeded out. As Stanley warned Eugene in *Brighton Beach*

Memoirs: Marry one of your cousins and you'll get babies with nine heads. In addition, the lack of diversity in immunity genes would compromise the child's ability to fight infection.

In 2008, a remarkable case of twin-cest played out IRL that illustrates both positive assortative mating and the incest taboo. Imagine finding that perfect someone—someone just like you—and then learning you are brother and sister. Much like Luke Skywalker and Princess Leia, this actually happened to fraternal twins in Britain who were separated at birth and raised by different families. After their wedding, they made the shocking discovery that they were siblings and immediately got an annulment.

Why Young Love Feels So Different From Old Love

Young love is a little like riding the Tower of Terror at Disney World. The fall is terrifying and exhilarating. You feel excited, but you might hurl. The diorama of your life has been shaken, and someone has replaced you on the center stage. You know these oddly wonderful and maddening feelings will stabilize, but you're not sure you want them to. What exactly is happening inside your hot mess of a brain as you fall in and out of love?

Because mating is a top priority for our selfish genes, they built a brain that loves love. When love comes to town (in other words, when you find a deck of genes that would be good to shuffle into your own), your neurotransmitters and hormones fluctuate wildly. In 2005 at Rutgers, anthropologist Helen Fisher, who literally wrote the book on love, conducted brain scans of people who were head over heels. Areas of the brain that are most active when lovers are thinking about one another are reward centers involving dopamine. Some of the same brain regions involved in young love are also activated during cocaine use, meaning that Robert Palmer was not far off the mark with his song "Addicted to Love." The lure of that dopamine reward is so enticing that it mobilizes us into hot pursuit, compelling us to go to the ends of the Earth in an attempt to win the object of our affection. We have dopamine to thank for all of history's romantic poetry, art, plays, movies, and songs. And we have dopamine to blame for Rick Astley.

In addition to dopamine, people experience jolts of norepinephrine as romance achieves liftoff, which explains why you're such a train wreck during infatuation. Involved in the fight-or-flight response, norepinephrine causes your flushed cheeks, sweaty paws, fluttering heart, and insomnia. It may seem like a strange hormone to secrete at the dawn of a romance, but it helps to keep us alert and on our toes so we don't blow it with our newfound love. Given how precarious young love can be, you and your new squeeze also experience a spike in the stress hormone cortisol.

As dopamine and norepinephrine go up, the mood regulator serotonin goes down. The lowered serotonin in lovebirds explains their annoying obsession with one another. In a classic 1999 study, psychiatrist Donatella Marazziti at the University of Pisa found that serotonin levels in new couples claiming to be madly in love had plummeted to the low amounts seen in people with obsessive-compulsive disorder. The reduction in serotonin is why young lovers just call to say "I love you" a thousand times a day. Because serotonin is also the precursor for the sleep hormone melatonin, its drop may be why those young lovers next door shook each other all night long.

In short, young love turns us into stressed-out, obsessive-compulsive addicts who can't get to sleep. But love not only alters our body chemistry like bad medicine, it also causes changes in the brain. Love feels like you're under a spell because your brain literally isn't thinking clearly. Brain imaging studies show that having a crush on someone deactivates neural pathways involved in negative emotions including fear and social judgment, which diminish your ability to assess the person's character objectively. Love can make you blind in the sense that it causes your brain to suspend ties to its analytical processes to foster the feeling of oneness between you and your heart's desire. To outsiders, it appears that you've taken leave of your senses, and in neurological terms that is exactly what happened. It's like you're wearing sunglasses at night.

Love moves pretty fast. The changes in your body's chemistry are rapid, and underlie the strange things you are doing with, to, and for your brand-new lover. In some ways, you never want this euphoric yet exhausting cocktail of chemicals to change. But on other days, you honestly wonder how long you can keep it up. Just as a sprinter reaches her limit, young love can't go on forever. Evolution had to build in a mechanism to douse the flames of passion, because it is not healthy to maintain high levels of cortisol and low levels of serotonin. More importantly, our bodies need to return to baseline to redirect our energy on raising the impending child our subconscious brain believes was the purpose of all this fuss. Granted, some of us would rather bob for apples in a kettle of starved piranha than have kids. But our brain assumes a child is the goal and adjusts our body's biochemistry accordingly.

We mentioned how young love parallels addiction to drugs. Just as people with addictions can build up a tolerance to drugs, young lovers can build up a tolerance to one another. Over time, we become desensitized to the dopamine rush that used to be triggered by the sight of your lover's cheeks (whichever ones you prefer to envision here). The excess norepinephrine and cortisol begin to subside, zapping a lot of the energy you put into the frenzied courtship, and your rational circuits come back online. You don't bring me flowers anymore . . . because we need to save money to buy diapers.

Other hormonal changes explain how our love for one another changes over time. In both men and women, testosterone is the key hormone fueling the sexual desire in a burning heart. In men, it is highest in the early 20s; women usually have testosterone peaks during ovulation. One reason why passion dies down is that there is a waning in these biochemicals as both sexes make less testosterone with age. Dopamine also gets further diminished as we grow more familiar with one another, which is why some people seek out a new partner or one-night love affair. But before you respond to that online profile and admit to liking piña coladas, you and your partner can try to rev up your dopamine engines by spicing things up with novelty.

Medicines that we take can also alter our body's biochemistry in ways that can quash romance. Drugs that raise serotonin levels like selective serotonin reuptake inhibitor (SSRI) antidepressants may thwart young love's need to lower serotonin. SSRIs may not only make it harder for someone to fall in love, but may also trick us into thinking that we don't love our partner anymore. SSRIs are known to blunt emotional responses and create feelings of indifference, which can adversely affect someone's affection for their loved one.

It would behoove us to get rid of the false impression that white-hot passion is supposed to burn an eternal flame. At the start, love rocks us like a hurricane, but eventually—mercifully—the tempest settles and we should enjoy sailing on tranquil waters. It's not unusual nor something to fret about, as lust ultimately loses to love in cultures all over the world. However, love can win in the end for those who are willing to cultivate a deeply satisfying relationship.

Are We Monogamous?

In 1987, pop star George Michael whipped conservatives into a frenzy with his song "I Want Your Sex." Exceedingly tame by today's standards, the song was akin to a devil's dirge back in the day, getting banned from many radio stations. How dare someone sing about wanting something that all living creatures must have? Michael insisted the song was about infusing lust into a loving relationship, and even used lipstick to write "explore monogamy" on a woman's back in the video. I was only a teenager at the time and assumed monogamy was some sort of daring sexual position.

Monogamy is the exception to the rule in the animal kingdom. Even in mammals, only about 3 percent pair up to rear their young together. People can be monogamous with only one partner for life, but it's no secret that we suck at it. The vast majority of people have had more than one sexual partner in their lifetime. According to the National Survey of Family Growth taken between 2002 and 2015, men average six sexual partners in a lifetime and women average four. The divorce rate in the United States hovers around 40 percent, and it is even higher for those who get remarried. It makes you wonder why we even try. That is, until you consider the advantages monogamy brings.

For most animals, their kids come out of the box more or less ready to go. But our kids don't stand a chance of surviving on their own for years. (I know some in their 30s that still live in their mother's basement.)

Our ancestors started to explore monogamy because it helped ensure the survival of our helpless infants. Supporting this idea is the fact that 90 percent of bird species stay together to work as a team, too. Eggs must be incubated 24/7, so one bird has to incubate the eggs while the other gets some grub, at which point they switch so the hungry parent can go eat. If their offspring is high maintenance, there is a higher likelihood that the species practices monogamy. Another advantage is that monogamy minimizes exposure to sexually transmitted diseases, some of which can cause infertility, miscarriage, and birth defects. Finally, having multiple children with the same person creates siblings of different ages who can work together for the benefit of the family.

Despite these advantages, many couples have trouble staying together for the long haul. Perhaps "till death do us part" is too much to ask. In a 2010 study, anthropologist Justin Garcia at Binghamton University found a gene variant in the dopamine receptor DRD4 that might contribute to infidelity. Recall that DRD4 variants predispose individuals toward impulsivity and risktaking behavior. In the context of monogamy, folks with the DRD4 gene variant report more than a 50 percent increase in sexual infidelity.

In monogamous animals, such as gibbons, swans, and beavers, the males and females are the same size, partly because males do not have to compete for mates, and therefore larger, stronger males are not selected for by evolution. In polygamous animals, which take more than one mate, females are generally smaller than the males. In humans, men are typically bigger than women, so by this criterion we (and our ancestors) fit a polygamous mating profile.

In their book *The Myth of Monogamy,* David Barash and Judith Lipton argue that we are socially monogamous but not sexually monogamous. Meaning that most of us pair up to form loving and stable relationships that last a long time (socially monogamous) but, like virtually every other animal on the planet, we tend to seek out a part-time lover (not sexually monogamous). Although some are content to strive for a purely monogamous relationship, others are trying consensual open relationships (that is, having a glass of milk elsewhere despite having a cow at home). A 2017 study by psychologist Terri Conley at the University of Michigan found few differences in relationship functioning between individuals engaged in monogamy and those in consensual open relationships. Contrary to popular belief, the study also showed that couples in the latter maintained more satisfaction, trust, commitment, and passion for their main squeeze than their booty call.

In her studies of divorce, Helen Fisher has noted that couples around the world tend to part ways around the fourth year of marriage, in their middle 20s, and/or with a single dependent child. In the handful of other mammalian species that pair up to rear young, along with most birds, a phenomenon occurs known as serial monogamy. The mating pair stick together just long enough to see that the babies can make it on their own (or that the mother is able to handle things by herself), then they go their separate ways. Fisher proposes that our hominid ancestors, much like some of today's existing hunter-gatherer tribes, typically had babies in four-year intervals. After four years, most women are able to finish taking care of the child until he or she is independent. The cracks appearing in the glue of monogamy among today's couples who have been married with children for four years or so may be an evolutionary holdover. Humans can be monogamous for life, but it may have been more common in our past to have been serially monogamous: pairing up to rear children for several years and then either having another child with the same partner or moving on to pair with a new partner. Serial monogamy is still very common today, which is why divorce lawyers are never hurting for work.

Our proclivity toward serial monogamy could explain why many couples start to resent one another after a few years of marital bliss. Those idiosyncrasies that you once adored now irritate the hell out of you. The jokes that used to make you roar with laughter now make you roll your eyes. The sex that used to bring you to tears now bores you to tears. Could those selfish genes of yours be the reason why love's on the rocks? Could our bodies be programmed to send out subconscious messages that drive us away from a single partner to diversify our genetic portfolio? Can we (should we) fight those natural impulses?

There are meritorious reasons why couples should attempt to stay together for life, but science is revealing why many people are not well suited to do so. There is no one-size-fits-all solution for human bonding, so best we stop deluding ourselves that everyone should stay together forever. The definition of a successful marriage could expand to include those who remain loving and kind to one another whether under the same roof or not.

Why We Stay Together

Every species has its mating quirks when it comes to maximizing the chances of successful reproduction. Proving that love bites, the female black widow spider often devours her poor guy after the loving, supplying her with extra nourishment for the impending brood. In a species of flies known as midges, the males literally go out with a bang—after sex, his genitals break off to seal the female shut so no other males can inseminate his conquest. Like many other species, spiders and insects are born self-sufficient, so the male is dispensable after fertilization. However, in animals that must take care of their young, males have a bit more utility. Breathtakingly portrayed in the 2005 movie *March of the Penguins,* male emperor penguins incubate their mate's egg for over two months, keeping it at 100°F despite the subzero temperatures in Antarctica. He nearly starves in the process, losing almost half of his body weight waiting for the female to return.

If human babies could fend for themselves, there would be less reason for men and women to stay together. But even after nine long months of development in the womb, our babies are still born very premature, with zero chance of surviving on their own. Parents with bloodshot eyes and missing tufts of hair will affirm that taking care of our demanding rug rats is a full-time job. It really helps to have a partner you can depend on when the children cry. Parents that stay together and work as a team practice what scientists call pair-bonding. Pair-bonding requires cooperation, which poses an interesting riddle for evolution: How did selfish genes make survival machines willing to make sacrifices for others?

The biology of pair-bonding has been rather hard to study because the vast majority of species don't do this; luckily, researchers have identified two types of voles (adorable, hamsterlike rodents) that have revealed the molecular epoxy that binds us together. The vole species *Microtus ochrogaster* (prairie vole) forms monogamous pair-bonds, whereas *Microtus pennsylvanicus* (meadow vole) does not. Because these two species of voles are nearly genetically identical, scientists could not ask for a better model system to learn about the biological mechanism behind pair-bonding. Why are meadow voles promiscuous and prairie voles not?

Since the early 1990s, neuroscientist Thomas Insel has been conducting pioneering research on the key ingredients that keep prairie voles together: the hormones oxytocin and vasopressin. The pituitary gland produces these hormones, which act on multiple sites around the body in addition to the brain. For example, oxytocin, which means "swift birth," enables uterine contractions for childbirth and letting down of the milk for breast-feeding. Scientists also found that it motivates mothers to care for their newborn.

Can something as beautiful as the love a mother has for her child really be reduced to a chemical? Curious scientists wondered what would happen to virgin rats, which show no love for someone's squealing, needy pups, if you shoot some oxytocin into their brain? The result: They no longer act like a virgin; they act like mothers. Virgin rats drugged on oxytocin defend, groom, and snuggle foster pups that don't belong to them. Stunningly, another study showed that a mother rat's innate love for her children can be erased if she is given an agent that blocks oxytocin's action in the brain. In people, oxytocin appears to operate in the same fashion; for example, the higher a mother's oxytocin levels during the first trimester, the more likely she will engage in bonding activities with her baby. Even in fathers, a shot of oxytocin into his nose (which travels straight to the brain) causes him to play more attentively with his baby.

Oxytocin's effects can also cross species lines. Giving a new meaning to puppy love, oxytocin levels increase in both you and your dog when you pet your canine friend. The same happens when you and your significant other do some heavy petting. Orgasm produces a burst of oxytocin that is believed to foster attachment between the couple, leading some to call oxytocin the "love" or "cuddle" hormone. Could the oxytocin released during sex contribute to monogamous pair-bonding? It certainly seems to be the case for our prairie vole friends. Like some kind of biochemical cupid, oxytocin given to a female prairie vole will prompt her to pair-bond with a male that she hasn't even mated. If the release of oxytocin is blocked in the prairie vole, they no longer pair-bond. Without oxytocin, prairie vole sex becomes casual, as it is in their promiscuous meadow vole cousins.

How about the other way around? If we give these "cuddle hormones" to meadow voles, can we make them fall in love? Not without a little genetic engineering. Due to a genetic difference, it turns out meadow voles don't have enough of the receptors for these hormones in the right region of the brain, regions associated with reward and addiction. But in 2004, neurobiologist Larry Young at Emory University used a virus to deliver a vasopressin receptor gene into the reward centers of the brain, which then made the meadow voles behave more like prairie voles. The few species that explore monogamy do so because they express more attachment hormone receptors in the brain, which is governed by slight changes in the DNA sequence that regulates the gene. Remarkably, this simple mutation is all that is needed to lasso two hearts together.

Many consider ourselves lucky that biology has given us these groovy attachment hormones to experience the satisfaction of love and contentment that often follows. But there is a catch. As pair-bonding strengthens relationships between mates and their offspring, it inherently increases the protective instinct, promoting mistrust and dislike of outsiders who might pose a threat to the family or group. As Queen Cersei tells her fellow Lannisters in *Game of Thrones*, "Everyone who isn't us is an enemy."

Injection of vasopressin into the brain of a virgin male prairie vole makes him possessive of a nearby female, and like a dutiful mate, he will aggressively defend her space from strangers. The promiscuous voles do not exhibit this type of aggression, and neither do prairie vole males until they've had sex and pair-bonded to a female. The pair-bond also causes him to chase away other females, reminiscent of studies that suggest oxytocin works to keep men faithful to their wives. In 2013, psychiatrist René Hurlemann at the University of Bonn in Germany showed that the reward center in the brains of men doped up on oxytocin lights up more when viewing their partner's face rather than the faces of other attractive women. It is often said a woman can cast a spell over her man, and it would seem that oxytocin is the active ingredient in that potion.

The aggressive tendencies produced by oxytocin during pair-bonding foster positive feelings about family that can extend to your countrymen—but, alas, negative feelings about outsiders. In one unsettling study, men were given moral dilemma choice tasks: for example, being asked to review a list of names and choose who they would put on a lifeboat that had limited space. Men given oxytocin were more likely to rescue their countrymen and deny seats to foreign-sounding names; men who did not get oxytocin did not display this bias. It seems oxytocin can bring out both the best and the worst in us.

Findings like these make scientists cringe at the "love" hormone nickname for the same reason they cringe at gene nicknames: It is misleading to pigeonhole their function. Oxytocin and vasopressin are multitaskers in the body, and whether their effects on behavior are good or bad depends on the context. If the ethnocentrism studies on the "love hormone" were completed before the pair-bonding studies, oxytocin could have easily been dubbed the "racist hormone" instead.

Monogamy and pair-bonding are far from simple behaviors, and with so many gears in play, it is easy to see why there is such variation in the strength and length of a couple's tenure. Mutations in either the oxytocin or vasopressin genes (or their receptor genes) could alter the product, the amount, or when and where they are distributed in the brain. Indeed, several variants in the gene that makes the vasopressin receptor have been linked to wandering hearts.

Undoubtedly other regulators of attraction and attachment are yet to be discovered, and these hormones likely play together in intimate ways. High testosterone can reduce vasopressin and oxytocin, and men with above-average testosterone levels are more likely to stay single or have an affair. Finally, some reports show that epigenetic factors influence the expression of these hormones and their receptors, which suggests the environment can influence the durability of a pair-bond.

Why Are Some People Attracted to the Same Sex?

On the surface, homosexuality doesn't appear to make biological sense because it rebels against the imperative to procreate. Occurring in less than 10 percent of the population, history has long assumed that physical attraction to members of the same sex is an anomaly. But that assumption could not be more wrong, as homosexuality has been documented in more than 400 species to date.

Homosexuality is in the air: A variety of birds, such as the Laysan albatross, vultures, and pigeons, as well as insects like flour beetles and fruit flies, engage in same-sex pairings. In the sea of love, you'll find homosexual activity in whales. On land, homosexuality is seen from the African savanna to down on the farm. Nearly one in 10 rams are just not that into ewe, and would rather have sex with other rams instead. Homosexual behavior has also been reported in elephants, giraffes, hyenas, and lions, as well as in other primates such as our cousin, the bonobo. Bonobos are so into free love that they've been called "the hippie ape." Both males and females are bisexual and use sex as a greeting and conflict resolver (behaviors that would really tax the human resources department at your office). The point is that homosexuality is prevalent in the animal kingdom, but no one argues that other animals make a conscious choice to be that way.

Several hypotheses have been put forth to explain how homosexuality can be beneficial to a family or species at large. Most of these ideas revolve around the concept of kin selection, whereby we work to ensure the passage of our family's genes into subsequent generations. Because we share more genes with family than with strangers, there is a selfish tendency to look out for our own kind. Gay uncles and aunts help support and nurture the family tree. Another idea, postulated by the eminent sociobiologist E. O. Wilson, argues that homosexuality may serve as a means of population control, bringing the species into biological balance with the resources available in its environment. Another idea comes from more recent genetic findings that suggest homosexuality is a "trade-off trait." For example, certain genes in women help increase their fertility, but if these genes are expressed in a male, they predispose him toward homosexuality.

Scientists are closing in on the factors that may give rise to homosexuality. Studies of twins suggest a genetic component to same-sex attraction, although not much more than 20 percent. In 1993, geneticist Dean Hamer at the National Institutes of Health famously proclaimed that he discovered the "gay gene" by linking male homosexuality to a section of the X chromosome called Xq28. A much larger study by another group in 2015 confirmed Xq28, along with a region of chromosome 8, as strongly influencing male sexual orientation.

Exactly which gene(s) in these chromosomal regions are responsible and how they predispose the carrier toward homosexuality remains to be resolved. Genome-wide association studies are under way to compare the genomes of people who are homosexual and heterosexual, one of which not only found variation again on chromosome 8, but also identified a new candidate gene called SLITRK6. This gene is expressed in a brain region called the diencephalon, a part of which differs in size between people who are homosexual or heterosexual. Future studies must be performed to verify if the variant SLITRK6 gene leads to changes in brain structure that influence homosexuality.

Studies in mice have uncovered additional gene candidates that could influence sexual preference. In 2010, biologist Chankyu Park at Korea Advanced Institute of Science and Technology linked sexual preference to a gene called fucose mutarotase (which they abbreviated "FucM," perhaps as a dig to naysayers who refute a genetic component). When the FucM gene was deleted in female mice, they were attracted to female odors and preferred to mount females rather than males. Studies by neuroscientist Catherine Dulac at Harvard University have shown that disruption of another gene can cause female mice to act like males. Female mice lacking a gene called TRPC2, which is present in brain cells and aids in pheromone recognition, displayed typical sex-crazed male behavior—these females engaged in masculine courtship rituals, pelvic thrusting, and mounting of mates. These female mice also enjoyed burping loudly and watching football with one paw down their pants.

Based on the evidence available to date, it seems highly unlikely that a single gene will govern sexual orientation. Such a complex behavior is likely to be orchestrated by many genes as well as influences from the environment (particularly the prenatal environment a developing fetus experiences in the mother's womb).

Epigenetics provides an attractive explanation why the hunt for gay genes has proved elusive, in that such genes may be present but not active unless they receive a certain trigger from the intrauterine environment. Epigenetics may also explain why birth order exerts an influence on male sexuality: Each older brother a boy has increases his chances of being gay by a third. One hypothesis asserts that each male pregnancy prompts the mother to produce a stronger immune response to boy proteins, which then affects gene expression in subsequent male fetuses through epigenetic changes. Further supporting a role for epigenetics in sexual orientation, several groups have found differences in the distribution of DNA methylation marks in humans and animals exhibiting homosexual behavior. In 2015 neuroscientist Margaret McCarthy at the University of Maryland was able to give female rats the characteristics of a male rat brain by injecting drugs that inhibit DNA methylation. The epigenetic drug made the girls, although anatomically female, behave sexually like men.

It is now well established that gender identity is separate than a person's anatomical sex—what's below the belt doesn't matter if the head thinks it owns the opposite equipment. There is a critical period during fetal development when hormones shape the brain into a male or female form, or perhaps something in between those two ends of the spectrum.

Many factors can affect the type and amounts of hormones a fetus is exposed to during its residency in the womb. Males with a genetic condition called androgen insensitivity syndrome (AIS) lack a functional receptor for testosterone. Males with AIS develop female genitalia and are usually brought up as girls, despite being genetically male (XY), and they are attracted to men. This tells us that testosterone is needed to "masculinize" a prenatal brain; if that doesn't happen, the child will grow up to desire men. Similarly, girls who have a genetic condition called congenital adrenal hyperplasia (CAH) are exposed to unusually high levels of androgens like testosterone while in the womb, which masculinize her brain and increase the odds of lesbianism. Exposing a female fetus to drugs such as nicotine or amphetamines also increases the odds of her being born a lesbian. In addition, female rats that during pregnancy experience stress, which reduces testosterone in the womb, are more likely to have males that display homosexual behaviors. Hormone fluxes such as these are likely to affect sexual orientation by acting on transcription factors that alter gene expression or through fetal epigenetic programming.

Whatever the underlying genetics may be, the end result appears to induce hormonal shifts during pregnancy that affect how the brain is configured at birth. In other words, men who are gay have a brain that is structured more like a woman's, and women who are gay have a brain modeled more like a man's. Neuroscientist Simon LeVay at the Salk Institute has pioneered studies that are consistent with this prediction. A specific region in the brain called the interstitial nuclei of the anterior hypothalamus 3, or INAH3, is two to three times larger in men compared to women. In a 1991 study, LeVay found that this region in gay men was closer to the smaller size seen in women. Studies in rats conducted by other researchers confirm that damage to a male's INAH3 brain region changes partner preference.

The overwhelming evidence shows that same-sex attraction is not unlike the attraction heterosexuals experienced. Both have a biological basis and are programmed into the brain before birth based on a mix of genetics and environmental conditions, none of which the fetus chooses. Heterosexuals never reminisce about the day they went out for a long walk and decided to only be aroused by members of the opposite sex. The only choice people have in the arena of sexuality is whether they choose to treat those who are different with the respect, dignity, and equality they deserve. People are people.

Do We Have a Soul Mate?

According to a 2011 Marist Poll, nearly 75 percent of Americans believe in soul mates: the notion that there is only one person out there who can make you feel like you're walking on sunshine. The idea of a soul mate is the epitome of romance, drilled into our psyche the moment we hear "and they both lived happily ever after." Who doesn't want to be swept off their feet by a flawless partner? It happens to people all the time in movies, TV, and books, so why not you?

Because math. There are more than 7.5 billion people in the world. In his book *What If?: Serious Scientific Answers to Absurd Hypothetical Questions,* Randall Munroe crunched the numbers and calculated that we would need to live 10,000 lifetimes to find our soul mate. In other words, if there is only one person right for you in the world, your chance of finding that individual is about the same as finding a working paper towel dispenser at an airport bathroom.

What if we relax the exclusivity component? Drawing from the same equation used to calculate the number of possible civilized planets in our galaxy (52,000, in case you were curious), Joe Hanson of *It's Okay to Be Smart* computed that 871 soul mates are waiting for you in New York City alone. Better, but the odds of bumping into one of those 871 soul mates next time you visit the Big Apple are still slim to none.

Don't despair—belief in a soul mate turns out to be detrimental to our relationships. People banking on a soul mate invest too much time second-guessing their choice instead of working on said relationship. Research shows that couples who believe in soul mates struggle with conflict more than those who view their time together as a journey with opportunity for growth. Soul mate diehards experience more relationship anxiety and are less likely to be forgiving after a lover's quarrel. The belief in a soul mate instills an expectation of perfection in our partner—and, if that is not achieved, believers in soul mates give up too hastily, thinking they must not have found Mr. or Ms. Right.

The notion of a soul mate is fodder that we should ban from our kids' plates, along with other mental junk food like superstitions and the supernatural. Belief in a soul mate is lazy. If you want to cultivate a satisfying relationship, you need to be a constant gardener. The good news is that there are many people out there with which you can make it work, not just one.

Almost Paradise

Influenced by our genes, evolutionary history, culture, epigenetics, hormones, microbiota, and more, the rules of attraction are anything but simple. But rather than in the stars, the language of love is written in our biology. We feel a chemistry, the vast exchange of information occurring beneath the surface that signals to our brain whether a person has the right stuff. But be mindful of the brain's superficial tendencies; the primal brain judges a mate by its cover. We need to exercise the logic circuits in our brain to determine if we want to become a part of that person's story—don't let the part of your brain going gaga over a flashy tail display win the day.

If you're interested in an everlasting love, science is revealing the secrets to turning a honeymoon into a permanent vacation. Brain scans of couples in successful long-term relationships show heightened activity in regions associated with empathy and controlling one's emotions. We've learned that it's natural for passionate love to yield to compassionate love. So although you don't need to sweat the loss of passion's heat, you don't need to give up trying, either. Surveys of older couples who are still in love after decades together often cite humor, sex, and novelty. Keep feeling fascination, as these are the things that serve up fresh shots of dopamine. To sustain a dopamine fix as a couple, some go on high adventures with one another. Some bungee jump together. Some go white-water rafting. My wife and I brave the Costco jungle on a Saturday.

Forged from genetic selfishness, love has emerged as a powerful new force on the evolutionary landscape that seems to have no bounds. The always prescient Bertrand Russell wrote in the 1930s that "Love is able to break down the hard shell of the ego, since it is a form of biological cooperation in which the emotions of each are necessary to the fulfillment of the other's instinctive purposes." This dissolution of ego not only promotes a fruitful relationship with your significant other, but also helps improve the welfare of people all over the world.

In his book *The Expanding Circle*, philosopher Peter Singer makes the case that the better angels of our nature, such as love, altruism, cooperation, and sacrifice, arose from the biological imperative to protect our genetic legacy, but have blossomed into conscious drives that are expanding the circle of moral concern from immediate family to towns, nations, and ultimately the world.

All of this from a few mutations in DNA millions of years ago that facilitated pair-bonding? Pretty groovy.

» CHAPTER EIGHT «

MEET YOUR MIND

If the human brain were so simple that we could understand it, we would be so simple that we couldn't.

—Emerson M. Pugh,
The Biological Origin of Human Values

As marvelous as it is, our brain took a surprisingly long time to realize its splendor. Our ancestors' brains weren't quite sure what to make of the gelatinous goo in our heads. The ancient Egyptians thought the brain produced mucus, which is understandable considering its appearance and proximity to nose and throat.

The first reference to the brain appears in the Edwin Smith Papyrus, believed to be written in Egypt in 3000 B.C.; it mentions that the brain helps people walk like Egyptians. The author linked brain injury to paralysis of the opposite side of the body, an astute observation that was lost in the sands of time.

Years later, the Greek physician Hippocrates presciently postulated that the brain is the source of all joy and horror that we experience. His compatriot, the physician Galen, rose to fame as a surgeon to Roman gladiators, which gave him unique opportunities to study all sorts of gore, guts, and brain trauma. Galen's experience treating gladiator wounds led him to conclude the brain was essential for movement.

The great philosopher Aristotle had a different opinion. He believed the heart was the body's most important structure, because it is the first organ to appear in the embryo, at the center of the body, and it beats. Moreover, cessation of the heart's rhythmic beat marked the end of life's song. He insisted, therefore, that this impressive muscle must contain our thoughts and control the body, while the boring brain just cools an overheated heart. Aristotle's ideas won the day, and no one gave much thought to the brain for centuries to come.

But the advent of the Renaissance brought discoveries that changed the way we thought about our thoughts. In around 1485, Leonardo da Vinci finally disproved Aristotle's idea that the heart was king when he cut the spinal cord of frogs, which instantly rendered the heart powerless. The heart was nothing without the brain! The discovery of nerves extending from the brain to all other parts of the body suggested that the brain was the commander, and used these fibers as conduits to give orders. Around the same time, a fellow named Luigi Galvani found that electricity could stimulate the movement of leg muscles in dead frogs. (You really didn't want to be a frog during the Renaissance!)

In 1803, Galvani's nephew, Giovanni Aldini, freaked out a lot of people with a public demonstration involving the corpse of a freshly hanged criminal and a pair of conducting rods hooked up to a battery. Aldini touched the electrical rods to the corpse's mouth, ear, and, of course, his rectum. The electrical currents caused the limbs of the corpse to jerk, his jaw to drop, and his eye to pop open. Despite the morbid nature of these experiments, they proved that the brain and nervous system use electrical impulses to animate the body, dealing a blow to the long-held belief that spirits did so.

Today we understand that the brain is who we are. If new bodies could be made in a laboratory to prolong our lives, which organ would you want to transfer to retain the essence of you? Your spleen wouldn't be helpful, nor your gallbladder, or even your heart. (Sorry, Aristotle!) The organ you'd need to carry on as you is your brain. It contains all your memories, feelings, and beliefs, all of which color your personality and behavior. Injuries to any part of your body below the shoulders might ruin your day, but they won't fundamentally alter your persona in the way a serious brain injury could.

The brain's got it going on, but it also must be recognized as a biological product of evolution that is far from perfect. Truth is, it has issues, some of which could get us all killed.

What's in Your Brain

Proposed by neuroscientist Paul MacLean, the triune brain model is a popular way to conceptualize how our most important organ evolved and functions. To help explain this model, we will use an ice cream sundae metaphor.

Representing the bowl of our sundae are neurons composing the brain stem; they are in charge of basic involuntary operations such as breathing and heart rate, as well as reflexive or instinctual behaviors. Also known as the "reptilian brain," this region is present in all animals. The ice cream in our sundae is made up of neurons specialized in integrating and responding to sensory information. Known as the limbic system, these neurons release chemical messages that contribute to emotions, reward and motivation, and learning and memory. Finally, the whipped cream on our sundae is the neocortex. Present only in mammals, it allows abstract thought, language, planning, and the ability to read this book. The human cortex is huge, comprising more than 75 percent of the brain. It has analytical abilities that apply reason to the knee-jerk reactions voiced by its more primal regions. E. O. Wilson describes the brain stem, limbic system, and cortex as heartbeat, heartstrings, and heartless, respectively.

The triune model is a greatly simplified view of the inner workings of the brain. Although represented as distinct parts, these three regions are constantly communicating and working together, as exemplified by people with brain damage that has cut off communication between the limbic system (emotions and memories) and the cortex (analytics). Those stricken with this injury can no longer make decisions or value judgments; they will deliberate in the cereal aisle all day without being able to choose what to buy. It appears that, unlike Spock, we cannot operate with logic alone; we must have access to our feelings and previous experiences to make decisions.

Also, many subsections of the brain reside within the three main parts. For example, the limbic system is composed of subsystems including the amygdala, hypothalamus, hippocampus, and cingulate cortex; the neocortex is further partitioned into lobes with discrete functions. But we're going to try and keep things as simple as possible here.

The neurons in your brain communicate by passing electrochemical signals to one another like a bucket brigade. Neurons do not touch one another, but have tiny gaps in between them called synapses. An activated, or electrified, neuron talks to its neighbors by pouring out buckets of chemicals called neurotransmitters into the synapses. The receiving neurons catch these neurotransmitters in buckets (receptors) perched on their surface. Astonishingly, each one of our billions of neurons can run up to 10,000 bucket brigades each. It would take 32 million years to count all the connections between the neurons in the average human brain, yet that's still not enough brainpower to solve most Sunday crossword puzzles.

Much of what we've learned about our most mysterious organ has come only in the last few decades, but we know enough to dispel some long-held myths. First, just like computers, a bigger brain does not necessarily mean a better brain. In fact, modern-day brains are 3 to 4 percent smaller than they were 15,000 years ago; what may be more important than brain mass is the number of connections between individual neurons. When you learn something, you are not adding brain cells but rather increasing connections between existing ones.

Second, we do not use just 10 percent of our brain (that is, unless you're watching *Jersey Shore*). It is a misconception that we possess unused regions that would give us mental superpowers if only we could tap into them, as in the 2014 movie *Lucy*. So the next time someone mentions this myth, ask them if they'd be cool with having 90 percent of their brain removed.

Third, your brain stinks at multitasking, so stop playing with your smartphone while driving and turn off email notifications while working (studies have shown that birds are actually better at multitasking!). Other research has shown that women have a modest edge over men in this arena. But this advantage disappears post-menopause, suggesting that the hormone estrogen is helpful for multitasking.

Finally, those popular "brain-boosting" activities, music, and games are not going to help you become a mastermind. Like with any training exercise, you might get better at the activity, but there is no evidence that those specialized skills translate into improved intelligence. Even the "Mozart effect," the 1990s craze that inspired parents to play classical music for their infants with the hope that it would create baby Einsteins, has been decisively debunked. According to science, the best things you can do for brain health is to eat right, exercise, and get a good night's sleep. Boring life hacks, but truth.

Still, here's another proven brain-enhancing activity that might come as more of a surprise to you: socializing. The social brain hypothesis contends that our brains evolved to be so advanced because we had to interact productively with large groups of people. Our ancestors not only had to fend off the usual threats of predators, famine, and weather, but they also had to navigate through thickets of rumor, heresy, and gossip to advance in the hierarchy of their society and recruit the best mates possible. As it was in the days of yore, socializing remains a good mental exercise, so don't delay in telling all of your friends about this wonderful book you're reading.

Oh, and one final myth to bust: Your brain is a stunning organ, but it's still a work in progress that suffers from some serious evolutionary growing pains.

Why Your Brain Has Issues

The brain is kind of a diva. It demands boatloads of attention, loves itself, can't admit to making a mistake, and even believes it's immortal. Let's start with the attention hogging. Of all the organs in your body, the brain demands the most energy. It guzzles 20 percent of your body's fuel, even if all you're doing is reading *People* magazine. To its credit, the brain has evolved ways to conserve energy, but these concessions have led to trade-offs in mental ability. For example, to save energy, your brain is often lazy and takes shortcuts. Instead of processing new sensory data all the time, it looks for patterns and makes assumptions. And it puts a lot of faith in its assumptions, to the point where it will double down on them even when presented with clear evidence to the contrary.

The brain gets squirmy when faced with uncertainty, like Linus without his security blanket. Uncertainty is at the heart of incomplete stories, which is why your favorite shows use cliff-hangers to keep you watching. The need for closure is so intense that your brain will plug the gaps in your knowledge with its own inventions. We do this all the time when attributing unexplained events or strange coincidences to religious or other supernatural forces. Rather than face the uncertainty of death, our self-important brain is convinced that we possess a soul that will live on after our body dies. It is a brilliant idea that has allowed us to focus on solving the problems at hand without being distracted by paralyzing existential questions.

Thanks to its runaway ego, our diva brain also believes it's better at doing things than it really is. People of low ability for a certain task tend to overestimate their skills in performing it. This is known as the Dunning-Kruger effect, named after the two psychologists at Cornell University who first described it in 1999. Their study, bearing the sobering title, "Unskilled and Unaware of It: How Difficulties in Recognizing One's Own Incompetence Lead to Inflated Self-Assessments," showed that participants scoring low on tests of humor, grammar, and logic totally overestimate their performance, just like a tone-deaf teenager on *American Idol* who thinks he's the next Bruno Mars.

The Dunning-Kruger effect can be amplified when drinking, which is why many people who say "Hold my beer" end up in the emergency room.

David Dunning and Justin Kruger were inspired to study this foible after learning about a bank robber who thought he could avoid being captured on video surveillance by applying lemon juice to his face. The dim-witted robber figured that because lemon juice can be used as invisible ink, no one would be able to see his face. This is akin to criminals who take selfies of their crime and post them to social media. Dunning and Kruger assert that people of low intellectual ability are too dumb to know that they are dumb. Even if your brain is an idiot, it will challenge geniuses and experts because, well, it is an idiot.

The Dunning-Kruger effect explains everything from why people think they can play bagpipes the first time they pick them up to why other people think climate change is bunk even though they've never read a science book. Do yourself (and the rest of us) a favor and (1) never play bagpipes, because no one should, and (2) keep your diva in check. As Confucius said, "Real knowledge is to know the extent of one's ignorance."

Why You Do Things for No Good Reason

When we are unwittingly compelled to behave certain ways due to our surroundings, we've been primed for that particular action. One way we can be coerced is through exposure to subliminal and supraliminal messages. Stimuli that we can taste, see, smell, touch, or hear are supraliminal. We are conscious of—but may not be focused on—them, like background music.

Stimuli that we are not conscious of are subliminal messages; for example, an image that flashes before our eyes so quickly that only the subconscious mind, not the conscious mind, perceives it. Studies show that these subtle messages may be responsible for some of our weird feelings that we can't explain.

Noted psychologist Robert Zajonc at the University of Southern California in Los Angeles showed that people will develop positive or negative feelings about foreign symbols if the symbols are preceded by a quick four-millisecond flash of a happy or angry face, respectively. Participants never recall seeing the face and can't explain their feelings toward the foreign symbols. Moreover, the emotions people attach to the symbols stick. If the experiment is repeated with the happy and angry faces switched, the subjects retain their original opinion about the symbol. These results support the assertion that a mind made up is resistant to change, presumably because the brain is rigged to conserve energy. (And let's face it: First impressions really are the most important!) In our everyday life, this means that we can form stable opinions about things without really knowing why.

Another freaky study, performed by a different group, tested whether we could be influenced by the logo of Apple, a company that is associated with exceptional creativity. After subliminal exposure to either the Apple logo or an IBM logo, subjects were asked to list how many uses they could imagine for a brick. Subjects who were flashed the Apple logo thought of far more uses for the brick than those flashed with the IBM logo. It seems an Apple a day may help keep writer's block at bay.

In addition to visuals, our actions can also be covertly influenced by sound, smells, temperature, taste, touch, or the kind of things we read. Wine shoppers buy more wine from a certain country if music from that country is being played in the store. We are much more likely to clean up our crumbs after eating in a room that smells like it has been freshly cleaned. We are inclined to think warmly of a stranger we meet while holding a warm drink, but think coolly of a stranger we meet while holding an iced drink. We're more likely to judge a questionable act as morally reprehensible if we've tasted something bitter. When we sit in a hard chair, we are more likely to drive a hard bargain (which is why car dealerships try to make us so comfortable). We will act more like a miserly Scrooge if we've read business magazines before playing a financial game.

Covert forces influencing our mood and behavior also emanate from our social media feeds. Facebook performed a controversial experiment in 2012, which intentionally manipulated a user's news feed by flooding it with biased posts. Results show that if our page is enriched with negative posts, we will post something negative. If our page is filled with positive posts, our posts will be positive.

The effects of subliminal priming do not affect everyone in the same way. Whether a subliminal message breaches our subconscious may depend on our personality. Sensation seekers, for example, are more susceptible to subliminal exposure to the Red Bull beverage brand (an energy drink that "gives you wings") than subjects who are not sensation seekers.

At one point, most of us have caught ourselves emulating characters we read about in books or see on film, a behavior that is even more relevant in our modern era of binge watching shows. After a night of watching one too many episodes of *It's Always Sunny in Philadelphia,* I strolled into a faculty meeting the next morning and very nearly addressed my esteemed colleagues with "What's up, bitches?" This subconscious adaptation is called "experience-taking," and it affects people differently because we typically mimic the characters we identify with at the time. When I was a kid watching *Fast Times at Ridgemont High*, I would identify with the students. Now I identify with Mr. Hand. Interestingly, if we read in a cubicle with a mirror, which serves as a reminder of our own self-identity, we are more immune to experience-taking.

The physical state of our body can also greatly affect our state of mind. We know to avoid coffee addicts until they've had their daily dose of caffeine, or else we will be dealing with someone oozing as much charm as *The Handmaid Tale*'s Aunt Lydia. We can become very impatient with someone who wants to talk about his weeklong vacation if we have a full bladder about to go supernova. "Hangry" is a term used to describe how quickly some of us jump to anger when we are in desperate need of a Snickers bar.

In most cases, we will be forgiven for being as irritating as our growling belly. But the hangry phenomenon can have serious repercussions, if, say, you're a prisoner up for parole. A 2011 study showed that a prisoner has a 65 percent chance of being paroled at the start of the day, but prisoners who are judged right before lunch are almost uniformly denied parole. After lunch, the chances of getting parole return to 65 percent. You may want to reschedule your court hearing, performance evaluation, or surgery for after lunch!

And you also might not want to wear black. As Professor Snape might gripe, most people have a subconscious bias when it comes to dark colors, which might stem from our ancient fear of the dark. Gregory Webster at the University of Florida in Gainesville has shown that sports teams wearing black jerseys receive significantly more penalties (presumably because the referees perceive the players as more aggressive). When the same team wears a lighter color for home games, this bias in penalties is not seen. The color red, long associated with fire and fury, also has real-world effects on people's perceptions. Athletes wearing red in boxing matches are judged to be more likely to win than athletes wearing blue or green apparel. Even the color of your medication can affect the way you feel. Patients perceive red and orange pills as stimulating and blue and green pills as sedative or calming, even when they are just placebos containing no active drug. Pills that are bright yellow make antidepressants more effective. As the deadline for this book came at me like a runaway train, I gave serious consideration to repainting my office to cover the soothing green walls with fire engine red.

People who believe they are being assaulted with subliminal messages are not completely paranoid. Our senses are bombarded with countless stimuli from advertisements every day—and, yes, these stimuli can affect our behavior. It is humbling, and perhaps disconcerting, to realize that we make many of our decisions while under the influence of an unrecognized stimulus. However, it is important to note that there is no strong evidence that subliminal messages can make you do things that you normally would not do. If you are not thirsty, a subliminal message for cola is not going to impact you. If you are not psychotic, the violence in entertainment will not affect you in any significant way.

Why Some People Are So Smart

Some people have no trouble keeping up with the millions of characters on *Games of Thrones*. And then, some can't even keep the characters in the movie *Cast Away* straight. Why are some brains like a lighthouse and most others like a flashlight?

Many twin studies have consistently shown that genes account for 50 to 80 percent of someone's intelligence (meaning that 20 to 50 percent is influenced by external forces like *The Magic School Bus* or *Beavis and Butt-Head)*. As such, the hunt for "smart" genes has commenced. In 2017, geneticist Danielle Posthuma at Vrije University in Amsterdam conducted an analysis of nearly 80,000 people and uncovered dozens of gene variants associated with intelligence, including many known to be active in the brain. Intriguingly, one of these genes, called SHANK3, makes a protein that fosters connections between neurons.

Earlier work from 1999 by neurobiologist Joseph Tsien at Princeton University showed that it is possible to make a mouse smarter by giving it more copies of a gene called NR2B. These mice were so much smarter than their normal companions that they were nicknamed "Doogie mice," after the boy-genius in the television show *Doogie Howser, M.D.* Doogie mice are better at learning and remembering, flying through mazes and puzzles way faster than normal mice.

NR2B encodes an NMDA (N-methyl-D-aspartate) receptor, the type of brain receptor that is disabled in some patients showing symptoms of demonic possession (see Chapter 6). It is thought that increased expression of this brain receptor increases communication between neurons. Earlier work by others showed that this receptor is activated quickly when young mice learn, but is slow to activate in older mice (which may be why it is hard to teach an old dog new tricks). In case you're getting the idea that we should engineer more of this gene into babies, be aware that there is a trade-off: Doogie mice have a more pronounced fear response than normal mice, possibly because they remember adverse events too well.

The environmental component of the intelligence equation begins at conception. Environmental pollutants, such as lead or nutrient deficiencies, can have serious harmful effects on brain development that diminish cognitive skills for life. It has been estimated that Americans have sacrificed 41 million IQ points from exposure to lead, mercury, and organophosphates in most pesticides. Pregnant women who consume drugs or alcohol, even in small quantities, put their unborn children at great risk of having permanent mental deficits. In 2017, a surprising study performed in rats showed that even fathers' drug use may lead to learning disabilities in their children, presumably by altering sperm DNA. Psychiatrist R. Christopher Pierce at the University of Pennsylvania showed that a male rat's cocaine use prior to mating was associated with epigenetic changes in the brain of his offspring. These epigenetic changes led to alterations in the expression of genes important for memory formation. In mouse models, maternal stress can also impact brain development in her child by altering the mother's vaginal microbiota.

According to studies by psychologist Robert Zajonc at the University of Southern California in Los Angeles, birth order can also affect intelligence. The IQ score decreases up to three points with each new child born into the family. The primary reason is believed to be related to the first children receiving more attention from the parents than the latter children. Some studies show that breast-feeding can lead to average IQ scores being eight points higher than babies who are not breast-fed; breast-feeding establishes a different gut microbiota, which is likely to affect brain development and function. And other research suggests that the time of year the parents decided to get busy can affect brain development: Children conceived during winter months show a higher rate of learning disabilities. Because our bodies require brief exposure to ultraviolet rays from sunlight to make vitamin D, researchers speculate that pregnant women may not be getting sufficient vitamin D for their unborn child during the winter.

Cultural differences and gender stereotypes also impact learning and performance. Despite multiple studies showing virtually no differences in mathematical ability between men and women, a stereotype still exists that men are better at math. Although there is no biological basis for it, the stereotype itself puts some women at a disadvantage. Women who were asked to identify their gender prior to taking a math test underperformed compared to those who were not asked about their gender. The stereotype adversely affects men as well, giving some an inflated sense of their math skills. Studies have also shown that more men pursue science and engineering jobs because they overestimate their mathematical prowess.

Researchers have found that learning environment can dramatically alter the motivation and performance of students with different achievement levels. High achievers function best when they believe a word test counts toward their grade; however, low achievers perform better if they believe the same word test is just a fun puzzle. Curiously, high achievers who think the word test is just for fun perform less well on it. Likewise, low achievers who think the test is for their grade perform worse. These findings argue that a one-size-fits-all style of education is doing a disservice to many students. Tailoring education toward a student's individual motivations and goals is more likely to produce positive results.

Although numerous forces that influence intelligence are beyond our control, the point is not to give up on learning. On the contrary, it appears vital that we recognize our ability to improve; otherwise we may become ensnared by a self-fulfilling prophecy. Psychologist Carol Dweck at Stanford University has shown that if you inform students that their intelligence is not fixed but can grow and improve, they perform better in school. Dweck has also shown that this "growth mind-set" can help offset the negative effects of poverty on academic achievement. Whether a lighthouse or flashlight, trying to make your brain shine brighter is always a worthy and rewarding endeavor.

Is There a Genius Sleeping Inside Your Brain?

In the 1988 movie *Rain Man,* Dustin Hoffman portrays the savant Kim Peek, who was born without a corpus callosum, the bundle of nerves connecting the brain's hemispheres. Peek never developed the fine motor skills needed to dress himself or brush his teeth, and he had a low IQ. But he would have crushed you in Trivial Pursuit with his encyclopedic knowledge. Nicknamed Kimputer, Kim had an off-the-charts photographic memory, able to recall mostly anything he'd read in books (reportedly 12,000 of them!) or heard in song. He was also a human GPS, having memorized road maps of all major cities throughout the United States in astonishing detail.

The otherworldly talents of savants vary. Ellen Boudreaux, born blind and autistic, can play a piece of music flawlessly after hearing it just once. Autistic savant Stephen Wiltshire is able to draw strikingly detailed landscapes from memory after viewing them for just a few seconds, earning him the nickname "Human Camera."

You may envy these superhuman abilities, but they usually come at a high cost. One area of the brain doesn't seem capable of thriving without pulling substantial resources from other areas. As we saw in *Rain Man,* nearly half of savants have autism-like characteristics and struggle with social interactions. In some savants, brain damage is so severe they can't walk or take care of themselves. Then again, Daniel Tammet is a high-functioning autistic savant who suffered epilepsy and behaves like any ordinary guy until he recites pi to 22,514 decimal places or speaks in one of the 11 languages he knows. Other human calculators, like the German mathematical wizard Rüdiger Gamm, do not appear to be savants with a brain anomaly. Gamm's gift is attributed to unidentified genetic mutation(s).

Perhaps even more fascinating are people who are leading perfectly normal lives and then acquire savant-like skills after some sort of head trauma. There are only 30 or so known cases in the world in which ordinary people suddenly obtain extraordinary talents following a concussion, stroke, or lightning strike. The newfound talent they acquire may be a photographic memory, musical abilities, mathematical genius, or artistic brilliance. It raises the intriguing question: What sort of hidden talent might be caged in your brain? If unleashed, would you be able to rap like Kayne West or bust a move like Michael Jackson? Math like Maryam Mirzakhani or paint happy little trees like Bob Ross?

Similarly, there's a curious connection between acquired artistic abilities and some forms of dementia like Alzheimer's disease. As neurodegenerative disease ravages the higher-order functions of the mind, extraordinary new talents in drawing or painting sometimes emerge. Another parallel between the appearance of new artistic abilities in people with Alzheimer's and savants is their single-minded focus on their talent at the expense of social and language skills. These cases lead some researchers to hypothesize that the destruction of brain regions involved with analytical thinking and language has allowed latent creative skills to shine.

Neuroscientist Allan Snyder at the University of Sydney in Australia has been working on a noninvasive method to temporarily quiet parts of the brain by shooting weak electrical currents through electrodes placed on the head. After he dampened activity of the same analytical region destroyed in people with Alzheimer's who transform into artists, his subjects were much better at solving a puzzle that required creative, out-of-the-box thinking. (I hate to break it to Snyder, who has spent thousands of dollars on his neurological equipment, but I can achieve the same result with a bottle of cheap wine.) At any rate, the findings led Snyder to believe that we all have savantlike abilities, but our brain deliberately suppresses them.

We are a long way from understanding if everyone has a little Rain Man chained up in the dungeon of their brain—and, if so, how to release these seemingly miraculous powers. But given the rarity of their appearance and the often debilitating trade-offs associated with these mad skills, I would not go banging my head against the wall in an attempt to turn them on just yet.

Why We Forget Things

In *Downton Abbey,* the butler Carson mused, "The business of life is the acquisition of memories. In the end that's all there is."

Our memories are indeed the most precious items our mind collects—not just for the sentimental value, but also for the survival value. We don't fully understand how memories are formed, much less how they are recalled (sometimes, many years after collecting dust on a shelf in our cerebral attic). But we have learned that memories, as airy as they may seem, are clearly a product of the material brain.

Currently, we think of memory in two stages: our short-term (working) memory and our long-term (stored) memory. Short-term memories are temporarily held until our brain decides they are important enough to convert into long-term storage, a transition that typically occurs during sleep. Repetition is one way to create a long-term memory, and structural changes can be seen in the brain as a consequence. Taxi drivers in London who memorized the labyrinth of streets through the city have a section of their hippocampus that is larger than average. Professional violinists (and probably my son and his gaming chums) have an enlarged area in their cortex that is associated with hand dexterity. It is believed that active neurons firing during these activities are releasing substances that facilitate connectivity and possibly growth in their surrounding area. It's like repetition summons a cerebral form of Hans and Franz, ready to pump *(clap)* your brain up.

When it comes time to recall a memory, our brain does not replay it like a video recorded on our phone. Our brain must reconstruct the memory every time it is recalled. The reconstruction process explains why our memory is imperfect, because different details may be incorporated each time we re-create the memory in our mind. Memory retrieval is a lot like the "Telephone" game, in which kids go around in a circle and whisper a short story into one another's ear. By the time the last kid hears the story, it has often deviated from the original. The core of the story usually remains the same, but the details have changed, sometimes substantially. Our memory is susceptible to the same trappings, which is why old friends at a high school reunion sometimes argue over details about the day they spent in detention together at the school library.

Memory retrieval can be affected by time and interference. Time shakes our mental Etch A Sketch; if memories are not redrawn on a more or less routine basis, most get fuzzy and eventually decay. Interference occurs when similar events are encoded on top of an existing memory, creating confusion between the two (a flaw that has been responsible for many breakups).

As most of us know from experience, memories that have a strong emotion attached to them are easier to recall, as are things we've heard over and over again. This is why I can sing along to almost any '80s song yet forget to pick up the raspberries my wife asked me to get at the store. When I really need to remember something, I try to incorporate it into the lyrics of an '80s song that I know well. For example, I'll modify Prince's "Raspberry Beret" to "She needs raspberries today, the kind you find in a grocery store." Problem solved.

Brain damage causes memory problems, which establishes that our memories are a purely neurological phenomenon. One of the most often cited examples is patient H. M., who in the 1950s had a portion of his temporal lobe removed to treat severe epilepsy. This radical treatment got his condition under control, but the bad news is that it left him incapable of forming new memories. Similar to Leonard, the protagonist in the film *Memento*, H. M. woke up feeling like every day was the time before his surgery. He could recall everything up to a decade before his surgery, but he could not form a new memory. He would greet his wife at noon and greet her again at 12:15 p.m., unable to remember that he already saw her. He died in 2008, still believing that Harry Truman was president. More common sources of brain damage that compromise memory include stroke, drug abuse, or diseases like Alzheimer's and Lewy body dementia.

In the absence of brain trauma, what excuses can you offer your spouse when you forget that she got her hair done? I'm not sure she'll buy it, but perhaps you can blame one of the genes proving to be important for memory. CREB is a transcription factor that regulates gene networks, and gene variants causing a gain or loss of CREB function improve or impair the formation of long-term memories, respectively. ZIF268 is another transcription factor important for memory, essential for the conversion of short-term memories to long-term ones. Other studies have linked variations in a gene called brain-derived neurotrophic factor (BDNF) to whether a person is good at remembering past events.

There is also evidence that epigenetics contributes to learning and memory, including instinctual behaviors. Like a new smartphone preloaded with apps, all living creatures are born with certain innate behaviors already programmed into their brain, whether a baby's cry, a bird's song, or a honeybee's dance. A 2017 study by neurobiologist Stephanie Biergans at the University in Queensland, Australia, took advantage of the complex dance moves honeybees use as a type of GPS to remember where food is located. By administering a drug that inhibits DNA methylation in the brain of the bees, Biergans was able to disrupt their memory skills. You may recall that DNA methylation is also linked to fetal programming in mouse pups born with a memory to fear cherry odors when their parent experienced a shock coinciding with the scent (Chapter 6). Collectively, studies like these have generated the provocative idea that instincts, and perhaps other learned behaviors, use epigenetic mechanisms to codify memories in neurons.

In addition to modification of DNA, the chemical modification of histone proteins associated with DNA is also critical for memory function. Histone acetylation, which activates gene expression, is required to increase expression of genes like CREB, BDNF, and ZIF268 during memory formation. A decrease in histone acetylation accompanies memory deterioration in disease or in the elderly. Drugs that inhibit histone deacetylase (HDAC) enzymes, which remove acetyl groups from histones, restore histone acetylation and improve memory in mice.

Even our microbiota could be affecting our memory skills. Germ-free mice lacking microbiota show impaired memory compared to normal mice possessing gut bacteria. In addition, studies by microbiologist Philip Sherman at the University of Toronto have shown that a bacterial infection in the gut of mice can disrupt the normal microbiota and cause persistent damage to learning and memory in the brain, even after the infection has been cleared. One way the microbiota could be influencing memory is through enhancing expression of BDNF in the hippocampus. Upon infection, BDNF levels drop; however, if Sherman fed probiotics to mice during the infection, they did not show a decrease in BDNF and were able to resist the learning and memory deficits caused by the infection.

If you're going to remember anything from this section, it should be that people are forgetful for many unexpected reasons—reasons that may not be their fault. At least we now have personal assistants on our smartphones to serve as an extension of our brain. We just have to remember to program them.

Why You Are an Illusion

Our brain is an extraordinary organ that appears to provide us with a newfound attribute that most creatures do not have: free will. When we think about whether to go dancing or stay in for the night, it truly feels like we are free agents in charge of making this decision. But could your subconscious arrive at this decision based on factors you're not aware of and then trick you into feeling like you've made the call?

The reason we even consider such a weird scenario is because of experiments Benjamin Libet started in the 1980s. Libet was measuring brain activity in people deciding when to lift their finger. Participants were instructed to tell Libet the instant they made their decision. The surprising result is that brain activity *precedes* our conscious awareness of the decision. These experiments have been refined with more sophisticated equipment, to the point where researchers can now review a brain scan and predict a person's decision up to 10 seconds before the person actually executes that decision. These experiments suggest that a lot is going on in our heads "behind the scenes" that we are simply not privy to until after the fact. The choices we make appear to be predetermined before we are consciously aware of them, and those voices in our head are echoes telling us what the brain has already decided, making it feel like we arrived at that decision. In other words, the movie has already been written and our sense of self is merely at the IMAX theater, feeling part of the action. In this sense, our decision-making process seems to be as involuntary as our heart rate and breathing.

As mentioned, a primary function of the brain is to simulate our reality to bring the world "out there" into our skull so our brain can respond. It would appear that what we call our "self" is just another character in this facsimile, and that our brain knows what we're going to do before we do. If this is true, our sense of self and free will are illusions.

Before your head spins right off your shoulders and you go to bed thinking that no one is accountable for anything, there is a school of thought that argues instead of free will, we have "free won't." Although we do not have control over the decisions our subconscious makes, some researchers argue that our conscious brain has veto power. In other words, our subconscious is like Congress drafting bills and our conscious brain is the president. We have no free will in which bills cross our desk. But we can decide which ones are sent to the recycle bin.

Free will or free won't, the truth is: Life doesn't require a brain, but a brain makes life worth living.

» CHAPTER NINE «

MEET YOUR BELIEFS

Never theorize before you have data.
Invariably, you end up twisting facts to suit theories
instead of theories to suit facts.

—Arthur Conan Doyle, *The Adventures of Sherlock Holmes*

As genes make proteins, our brains make ideas. Other brains critique these ideas, which fight to survive and reproduce. As with genes, useful ideas are expressed and ineffective ones are silenced. The more brains that support a particular idea, the more likely it will become a belief that permeates the culture.

Brains must work together to transform ideas into beliefs, or to retire beliefs that are outmoded. But when you ask a bunch of diva brains to collaborate, strange things happen. One would think that evaluating whether an idea is good or bad is a straightforward exercise in logic, as quantifiable measures for the idea's success should be attainable. However, when it comes to our beliefs, the majority of us are rarely objective and rational. Psychological experiments have revealed some intriguing, and sometimes depressing, things about how our brain mingles with other brains to shape human behavior and beliefs.

Let's begin by taking a look at why so many people find it difficult to speak up and raise hell against bad ideas.

Why Most of Us Are Not Rebels

It's fun to smirk into the mirror and pretend we're badass rebels bent on bucking the system and sticking it to the Man. We're going to show them who's boss, and take over the world . . . right after another episode of *Orange Is the New Black*. If we're being honest, the most rebellious the average person will ever get is sneaking 11 items into the 10-items-or-less checkout lane. We don't wear orange jumpsuits, but we are serving life sentences in an invisible cage, constrained by others who have a leg up on us. We have a long evolutionary history of not rocking the boat, a behavior that extends back to our prehuman ancestors.

Primate brains have been evolving ways to get along for millions of years, and have settled into a hierarchical structure based on rank. Chimpanzees are enmeshed in complex social webs and form alliances around different high-status individuals, much as we do. Predominant traits of alpha males in chimpanzee societies include physical strength, cunning, and the ability to recruit loyal friends. To challenge the alpha male could be a fatal mistake. This ingrained fear may partly explain why we hesitate to join a rebellion: If you rock the boat, you may find yourself walking the plank. So in terms of survival and reproduction, being a wallflower is an effective way to make another wallflower.

We also have a surprising predisposition to obey authority figures, whether they are parents, teachers, priests, police, Bruce Springsteen, or tribal leaders. From an evolutionary perspective, it makes good sense to listen to those who have more experience, as their knowledge is likely to help us survive and procreate. However, our receptiveness to authority has disturbing pitfalls as well.

Stanley Milgram at Yale University conducted a classic experiment in 1963 that put our obedience to the test, the results of which inspired Peter Gabriel's song "We Do What We're Told (Milgram's 37)." Milgram's study was prompted by the Nuremberg war criminal trials, in which the accused defended their heinous actions with a "just following orders" excuse. Could ordinary, nonviolent people hurt a stranger just because an authority told them to? Milgram devised an experiment in which his subjects (all men) believed they were helping students improve their learning skills. When the student gave a wrong answer, the authority figure in charge of the experiment (the experimenter) told the subject to press a button that would give the student a mild shock. Unbeknownst to the subject, the experimenter and the students were actors. As the student pretended to answer questions incorrectly, the experimenter stated that the magnitude of the shock must be increased. The student would wince in pain and, as the shocks got worse, would start howling in pain. Again, this was all an act. If the subject expressed doubt about inflicting more pain on the crying student, the experimenter reminded the subject how important it was to complete this experiment. Shockingly, two-thirds of the subjects continued to torture the student to the point where they were told that the strength of the shock was extremely severe, possibly lethal.

What were these subjects thinking when they delivered potentially life-threatening shocks to the learners? Perhaps they weren't thinking. Other researchers later found that brain activity is diminished when we follow orders. When coerced, our brain experiences a decreased sense of agency, meaning we feel less responsible for our actions.

The 2012 movie *Compliance,* which was based on real events, also portrays how easily we can be led into doing scandalous things if we believe an authority figure is issuing the commands. The movie depicts one of the many prank calls made to 70 fast-food restaurants between 1992 and 2004. Claiming to be a police officer, the caller was able to persuade the manager to detain an employee for an alleged theft. In some cases, the caller was able to convince the manager to perform a strip and cavity search on the innocent employee under the guise of helping law enforcement find the stolen items.

As a social species, we often conform our behavior to align with certain groups. Fitting into the group norm requires deindividuation, whereby individual members can lose sight of who they are and how they normally behave. Philip Zimbardo's famous 1971 Stanford prison experiment demonstrated how easily we fall prey to a group identity. Stanford students were placed in a mock prison and randomly assigned to pretend to be prisoners or guards. The experiment had to be stopped in less than a week, because the students pretending to be guards were abusing the pretend prisoners so badly that the latter attempted a failed rebellion, some even suffering depression and psychosomatic illnesses.

There are drawbacks to the prison experiment, including the small sample size and lack of rigorous repeats of the study. Additionally, some people interpret the results differently. Maria

Konnikova wrote, "The lesson of Stanford isn't that any random human being is capable of descending into sadism and tyranny. It's that certain institutions and environments demand those behaviors—and, perhaps, can change them." In other words, if we're predisposed to behave the way we're expected to behave, then we may be able to shape our actions in more productive ways by altering those expectations.

What these examples teach us is that our brains are highly susceptible to conforming to group expectations and obeying authority figures. The point is not that we should descend into anarchy, but to recognize that our innate tendency to obey and conform can be hijacked by people with less than noble intentions. Studies like these do not pardon pranksters or war criminals, but knowing our brain's vulnerabilities is empowering. We must be on guard and think for ourselves.

One day our brains may even realize that our social hierarchies are a pyramid scheme that benefit only a few at the expense of many.

How Groups Become Polarized

As our brain encounters other brains, it becomes apparent that some are on the same page as ours, whereas others are not even in the same library. True to its diva form, our brain enjoys surrounding itself with like-minded brains. This biased brain of ours helps explain why our political system is so frustrating. Seen every day in the halls of Congress, "groupthink" is a form of Darwinian peer pressure that assails reason and compromise. Groupthink occurs when two (or more) groups have differing opinions, with each member vying to advance in the hierarchy of their own group by exerting the most rigid and extreme version of their opinion. Members of the group with more moderate opinions find themselves following the extremists to conform, or they risk being ostracized by the increasingly radicalized group. Reason often takes a backseat to preserve intragroup harmony and loyalty. The end result: We form highly polarized groups driven by fanatical views that essentially have zero chance of reaching a compromise.

The result of groupthink is a loss for both parties, who now despise one another for the gridlock they had a hand in creating. Legislation that manages to get passed tends to be an extremist policy born of mob mentality, rather than thoughtful discourse between individuals. The chasm is so great between parties that we talk about legislation as a political "win" or "loss," as if governing millions of people is some kind of game. To fix this, we need to be wary of groupthink and the danger of conforming to extremist views. Those who have forgotten that we're all on the same team need to be sent home.

Psychologist Mark Levine at Lancaster University in the U.K. illustrated an example of how we can use this knowledge for good. To understand the experiment, you need to know that Manchester United (MU) and Liverpool Football Club (FC) are two rival soccer teams. Levine had MU fans complete a questionnaire about the team and its loyal fans. Then they were told to report to another building. As they walked to the other building, an actor posing as a jogger pretended to fall and cry out in pain. The jogger was in luck if he was wearing an MU shirt, as nearly every MU fan offered assistance. However, if the jogger wore a plain shirt with no team logo, only one-third of the MU fans would stop to help. More depressing, if the jogger was wearing an FC shirt, even fewer of the MU fans bothered to help the fallen man. The results reveal a disgraceful side of human nature (although your brain is probably insisting right now that you would have helped anyone regardless of the shirt he wore). Maybe so, but remember how easily your diva brain lies to itself and often makes you out to be a better person than human nature allows. What if that jogger was a member of the opposite political party, someone from a company that put yours out of business, someone your religion tells you is evil, or a Justin Bieber fan? Don't fret too much, as there is a happy ending. Levine ran the experiment again (with different MU fans), but gave them a questionnaire that got them thinking about the camaraderie of soccer fans in general, not just one specific team. This time, joggers wearing either an MU or FC shirt were in luck; those wearing plain shirts, not so much.

I view this experiment as a hopeful ending, because it shows that we can be decent to people outside a narrow alliance. If we constantly remind ourselves that we're all part of a larger group, we can rescue ourselves from the jaws of polarized politics. And if we have any sense at all, we will extend our alliance to all fellow human beings on this pale blue dot.

Why Political Arguments Make You Want to Pull Out Your Hair (Or Their Hair)

Most people define themselves by their political beliefs, whether they lean to the conservative right or the progressive left. We like to think that we've arrived at our political views through our objectivity and critical thinking skills. We are deeply vested in our stance on the hot-button issues of the day, and can't fathom why the other side doesn't see things the way we do (which is, of course, the correct way). Can science shed some light on this quagmire? Could there be a biological basis for conservatism and liberalism?

As we've seen in many other cases, certain genetic variants predispose people toward certain personality traits, and political leanings are no exception. Studies show that political beliefs are more closely aligned between identical twins than fraternal twins, supporting the idea that genetic factors influence what we do at the polls, what sort of stickers we slap onto our bumper, and what news network we prefer. Remarkably, identical twins separated at birth and raised in different environments still found themselves in agreement on political issues when reunited. Political scientist James Fowler of the University of California in San Diego calls the hunt for genes associated with political affiliation "genopolitics."

One gene, in particular, which has popped up in numerous studies, correlates well with how we rock the vote. You may be able to guess which one it is by now, as we've seen it a few times before: DRD4. You'll recall DRD4 encodes a receptor for dopamine, variants of which can produce "daredevil" behaviors that include a proclivity for exploration, experimentation, and novelty seeking. As you might surmise, progressive liberals more commonly possess the daredevil variant of DRD4 than conservatives. Our genes also contribute to how our brain is built and, as we'll see in the following, neurologists have noted interesting differences between the brains of conservatives and liberals. It would seem that we may already be tilted toward one end of the political spectrum before we see our first campaign poster. But the latest polls say that genes alone cannot be the only factor involved. There are Republicans who ski the K12 and Democrats who never want to leave their house.

One of the best ways to test whether our political affiliations may be something we are born with would be to assess the personality of very young children and then ask if they are Republican or Democrat decades later. Lucky us: That experiment has been done! Psychologists Jack and Jeanne Block at the University of California, Berkeley, administered personality tests to nursery school kids and then tracked them down 20 years later to ask them politically loaded questions. Their findings revealed some signature personality traits in toddlers that strongly correlated with that person's future political affiliation: "The relatively liberal young men, when in nursery school two decades earlier, impressed nursery school teachers as boys who were: resourceful and initializing, autonomous, proud of their blossoming accomplishments, confident and self-involving. The relatively conservative young men, when young boys, were viewed in nursery school as: visibly deviant, feeling unworthy and therefore ready to feel guilty, easily offended, anxious when confronted by uncertainties, distrustful of others, ruminative, and rigidifying when under stress."

In regard to women, the study found: "Relatively liberal young women . . . 20 years earlier were evaluated in nursery school by a coherent host of qualities: self-assertiveness, talkativeness, curiosity, openness in expressing negative feelings and in teasing, bright, competitive, and as having high standards. The relatively conservative young women, as young girls in nursery school two decades earlier, impressed the then assessors as: indecisive and vacillating, easily victimized, inhibited, tearful, self-unrevealing, adult-seeking, shy, neat, compliant, anxious when confronted by ambiguity, and fearful."

There are also fundamental differences in how members of each political party behave. Who do you think is more likely to wear a starched shirt versus a Grateful Dead T-shirt? Studies have shown that conservative students are more likely to have an ironing board, a flag, sports posters, and a tidy dorm, whereas liberal students have piles of books, maps of the world, a wide variety of music, and a less orderly dorm. Personality profiling shows that, in general, liberals are more open-minded, creative, curious, and novelty seeking, whereas conservatives are more orderly, conventional, and better organized. In broad terms, liberals favor change when presented with new evidence, whereas conservatives favor stability guided by tradition. Not surprisingly, this is why liberals tend to be skeptics and conservatives tend to be religious.

Studies by political scientist John Hibbing at the University of Nebraska suggest that conservatives and liberals react differently to unpleasant images or startling sounds, with conservatives exhibiting stronger physiological reactions to the noxious stimuli. Subjects who were jumpier at threatening sounds and images favored defense spending, capital punishment, patriotism, and war, whereas the less ruffled subjects favored foreign aid, liberal immigration policies, pacifism, and gun control.

It follows, then, that just as most conservative talking heads who deliver blustering apocalyptic broadcasts, their listeners tend to be more paranoid due to this overactive fear response. Liberal talk shows are a yawn by comparison, as liberals are generally more levelheaded and calm, a trait that can backfire if a genuine threat is underestimated. Liberals generally tolerate uncertainty better and appreciate the complexity of issues whereas conservatives generally make fast and "low-effort" decisions, because they view the world more simply, more black-and-white. One method is not necessarily better than the other;

some situations call for a quick decision, other situations should be addressed with a more thoughtful approach. Ideally, we should work with humble and honest integrity to balance the two approaches. But who would watch that over an insult-laden grudge match?

The different ways liberals and conservatives view the world have real consequences on public health and policy, as illustrated in a 2017 study that examined how Democrats and Republicans felt about the causes of our nation's obesity epidemic. Republicans tend to blame victims for poor lifestyle choices and feeble willpower, whereas Democrats see greater complexity in the issue and acknowledge genes as a component of the problem. Consequently, Republicans are more likely to disfavor government intervention to help people with obesity, and Democrats are more likely to favor things like sugar and soda taxes as junk food deterrents. Additional studies confirm that conservatives would rather address problems through applying broad restraints to avoid a potential negative outcome, whereas liberals would rather use targeted interventions in hopes of encouraging a positive outcome.

Fresh off the heels of his Oscar nomination for playing King George VI in *The King's Speech*, actor Colin Firth issued a challenge to scientists in 2010, asking them to figure out what was "biologically wrong" with people who disagreed with him on hot-button political issues. Neuroscientist Geraint Rees at University College London accepted this challenge by examining the brain structure of liberals and conservatives, and granted Firth co-authorship on the study.

The structural patterns Rees discovered were so consistent that researchers could soon predict with 72 percent accuracy which party participants supported just by looking at their brain scan. Although conservatives tend to have a bigger amygdala, the brain region that is activated during fear and anxiety, liberals have a bigger anterior cingulate cortex, a region involved in critically analyzing instinctive thoughts. It would be interesting to look at the brain structure and activity in moderates or independents; one might expect their amygdala and anterior cingulate cortex to be in proportion and equally active, perhaps achieving a better balance of fear and rationality. Intriguingly, a larger amygdala has also been found in procrastinators, which may explain why they keep putting things off—they are frightful that their actions may produce a negative result. In a political context, this may help explain the different attitudes conservatives and liberals have for new ideas that shake up the status quo or break tradition.

Now we know why those clever memes we post haven't persuaded our Facebook friends on the other side to come around. You're not asking them to merely change their mind; you're asking them to change their brain. These findings indicate that the way our brain is structured plays a significant role in how we react to potential threats, stress, and conflict, which in turn correlates with our political affiliation.

There is a sound evolutionary reason why our species contains a mix of individuals along the political spectrum: Conservatives are excellent at detecting possible threats, and liberals are excellent at threat assessment. In a collaborative society, these complementary skill sets provide a prudent means for civilization to progress. The problem today is that we no longer respect one another's talents because we fail to rebel against the extremists who turn us on one another. It is far easier to dismiss someone as a "conspiritard" or "libtard" than to break rank and lend a thoughtful ear to the other side's concerns. We must break our addiction to the short-lived dopamine reward delivered by tribal groupthink and strive for the long-term reward that comes from using logic and reason to reach a compromise. In 1991, the rock band Live warned us about living in a black-and-white world; it is time we learn to see "the beauty of gray."

Why It Is Hard—but Not Impossible—to Change Their Mind

When it comes to changing minds, you'll have a better chance at affecting younger ones. The brain is not finished maturing until our mid-20s, at which point it can become stubbornly resistant to change on certain issues, like lava that has solidified. Why might our cognitive fortress be so impenetrable at this point? Sometimes not even mountainous evidence and Vulcanesque logic can shake demonstrably false beliefs.

Brain imaging is providing new, but disheartening, insights into the way we think about the beliefs we hold dear. In one study of politically liberal people, researchers presented subjects with several political statements (for example, "Abortion should be legal") and several nonpolitical statements (for example, "Taking a daily multivitamin improves one's health"). Participants stated whether they agreed with each statement, then were presented with counterarguments for each statement.

The results were intriguing. Participants generally had no problem reevaluating their stance on the nonpolitical statements, but would not budge on the political ones. When political beliefs were challenged, subjects showed more activity in their amygdala, as if a threat were being perceived. As with other threats, emotion goes charging like cavalry to invade the decision-making process. Additionally, regions of the brain associated with self-representation lit up when political beliefs were challenged, suggesting that our brains have serious trouble dissociating those beliefs from self-image. In other words, challenges to our political stance or our party's leaders challenge our identity. Our diva brain will have none of that! Rather than face an identity crisis, we will deny evidence or give it the stink eye, just like any good press secretary.

We get a dopamine reward whenever we hear someone is in agreement with us, so it is not surprising that we seek out arguments that support our beliefs, like ET following a trail of Reese's Pieces. Fortunately (or unfortunately), the Internet makes it easy to find someone out there who shares our beliefs, no matter how nutty they may be. Our diva brain raises a toast to evidence that supports its beliefs, but throws a drink in the face of evidence to the contrary. This unseemly behavior is known as confirmation bias.

A 2016 study done by neuroscientist Tali Sharot at University College London illustrates how confirmation bias is like willingly poking out your mind's eye. Sharot divided participants into two groups, based on whether they believed that human activity was accelerating climate change. She then told some people in each group that scientists reevaluated the data and discovered that climate change was happening even faster than originally thought. She told the others in each group that the analysis showed climate change was actually not as bad as originally thought. Guess who believed what. The people who believed in climate change scoffed at the new analysis that concluded it was not so bad, but those who were told it was getting worse accepted the new analysis. The opposite was true in the climate change deniers. Deniers who were told the problem is even worse refused to accept the analysis, but deniers who were told the problem is not so bad accepted that analysis. In other words, our brain has a disturbing tendency to only consider evidence that reinforces its current beliefs. Like most of us, the people in this study are not living by Thomas Huxley's words of wisdom: "My business is to teach my aspirations to confirm themselves to fact, not to try and make facts harmonize with my aspirations."

Confirmation bias is a nasty habit that seems to defeat the purpose of having a brain in the first place. But it persists because the emotional parts of our brain evolved first and have been around much longer than our newer capacity to reason. This may be why emotion still frequently wins the day over logic. Psychologist Drew Westen at Emory University put confirmation bias under the microscope (more precisely, into a brain scanner), and found that the analytical part of our brain was deathly silent when subjects were presented with clear examples showing their political party leaders contradicting themselves. The parts of the brain that were active involved emotional responses. Westen also observed that when subjects were told something positive about their preferred politician, the reward center of the brain released a lot of celebratory confetti. As summarized by Westen, "Essentially, it appears as if partisans twirl the cognitive kaleidoscope until they get the conclusions they want." Data that make the diva look bad are dismissed as fake news. The same results occur in both liberals and conservatives, so at least they have that in common. Confirmation bias is why your best arguments fall on deaf ears, and why their arguments bounce unheard out of yours.

It seems hopeless, but high school teachers have a potential solution they use in debate club. Instead of defending your side of the issue, try defending the opposing side. As we begin to acknowledge the meritorious claims on each side, we may be able to attain a healthier dialogue that crawls toward compromise. Acknowledging that evidence usually fails to convince people who have already made up their mind, Tali Sharot advocates an approach that taps into a person's emotions, curiosity, and their power to solve a problem. For instance, anti-vaxxers who still believe the fraudulent study linking vaccines to autism are notoriously resistant to the hundreds of studies showing no association. But when reminded of the potentially devastating consequences of measles, mumps, and rubella, three times as many change their attitude toward vaccination. A more productive exchange may be possible by diverting focus from the point of disagreement to a common goal.

Why We're Religious

Almost as mystical as religion itself is the fact that most people on the planet are religious. Like language, trade, tool usage, and Starbucks, religion of some kind is seen in every culture across the globe. Whether you believe in a single male god with a billowing white beard, the alien leader Xenu of the Galactic Confederacy, the Force, or the disciples of a fire-breathing snail, the common theme among religious people is faith in the unseen. Despite being born atheists, most of us seem unusually receptive to religious indoctrination, as if we're built with a socket that spiritual appliances can plug into.

Part of the reason why we readily adopt a religion with no questions asked relates to our instinct to obey authority figures like our parents. The fact that many of us grew up believing in the incredulous story of Santa Claus showcases how blindly we accept what parents tell us. Religious ideas may be more enduring because our brain is squeamish in the face of uncertainty—for example, what happens when we die. As long as we keep getting presents, we can handle a world without Santa. But when a brain contemplates its final curtain call, that is enough to send any diva over the edge. Life moving on without me?! No way! I am too important! I will live forever! The prospect that an eternal spirit within us survives the demise of our body is enticing and distracts us from dwelling on the likelihood that our existence is a finite blip on the radar. As the philosopher Albert Camus observed, "Humans are creatures who spend their lives trying to convince themselves that their existence is not absurd." Religion is chicken soup for the diva mind.

Throughout history, many people who have taken a blow to the head, eaten a strange mushroom, had a fantastical dream, or suffered epileptic seizures may have been convinced there is something more than the plane of consciousness we inhabit. Such experiences fuel the belief in a spiritual world, although there's no tangible evidence of such a realm outside our imagination. Recent studies of the brain validate that our sensations of spirituality are literally all in our head. Out-of-body and other spiritual experiences can be triggered artificially just by tickling the brain with an electrode or by taking hallucinogenic drugs that get the brain tripping. Often cited as evidence of an afterlife, near-death experiences have also been invalidated because the brain continues to operate for a brief time after we flatline. A study in rats showed that a surge of neuronal connectivity in a dying brain exceeds the amount seen during normal conscious states. Similar surges in the brain's electrical activity have been recorded in dying human patients as well. What this means is that the mammalian brain experiences heightened consciousness as the brain is dying, providing a likely explanation why people may have vivid spiritual experiences or a feeling that they've come back from the "other side."

To put out-of-body experiences to the test, a clever 2014 study placed objects on the top shelves in hospital resuscitation rooms. These objects would stick out like a sore thumb but could only be seen from a bird's-eye view, so if a revived patient claimed to be floating above the room, researchers could ask if they saw the strange object. Such patients typically describe the usual things one associates with a hospital room (doctors, nurses, medical equipment), but none in this study reported seeing an unusual object. (Disappointingly, the patient who described his surroundings most accurately was revived in a room that did not have an unusual object placed in it beforehand.)

Our brain is vulnerable to generating images that could easily be misconstrued as otherworldly or spiritual, so it is understandable why so many people embrace religious ideas.

And embrace them we have! There are more than 4,000 religions in the world, each convinced of its authenticity with equal veracity. It's very telling that the overwhelming majority of us stick with the religion our parents taught us. For most of us, our religious persuasion wasn't a choice any more than our native tongue. You can argue you have the right to change religions, although some faiths will excommunicate or even kill you for browsing the global theology market. But assuming you are granted the right to shop around, receiving religion in childhood is like getting a tattoo on your brain, and it is exceedingly difficult and painful to erase.

In addition to quelling existential crises, religion offered an evolutionary advantage in terms of saving brain energy. As author Edward Abbey wrote in *A Voice Crying in the Wilderness,* "Whatever we cannot understand easily we call God; this saves much wear and tear on the brain tissues." Like duct tape, religion provides a quick fix for problems we have little hope of solving in the short-term, such as where we come from, the meaning of life, and what happens when we die. Religion patches up those pesky holes in our knowledge, freeing our brain to focus on more immediate problems of the day that pertain to our primary duties to survive and reproduce.

But as humans developed agriculture and laborsaving technologies, they found time to ponder bigger questions in life besides what's on the dinner plate and who to mate. Some started fiddling with the duct tape to see if they could figure out better solutions. Religion has grudgingly surrendered ground since the Enlightenment, effectively ending that grim period in human history called the Dark Ages.

We have found astounding truths to replace the duct tape. God did not create all living creatures as is; evolution sculpted them through natural selection. God does not make the day suddenly go dark; that is caused by a solar eclipse. God does not make the Earth shake; earthquakes are the "fault" of underground rocks breaking and releasing seismic waves. God does not cause leprosy; it is caused by bacteria called *Mycobacterium leprae*. And so on.

Despite the monumental discoveries driven by science, some people still really love the duct tape. They react as if the duct tape is attached to their own hairy skin. The agony felt when replacing the duct tape is cognitive dissonance, a psychological term that refers to conflicting knowledge that screws with your worldview. Your diva brain likes the world that it knows, and if a new discovery comes along that refutes one of your mind's most cherished beliefs, this creates cognitive dissonance. Remember the sheer disbelief Luke Skywalker expressed when Darth Vader revealed that he was Luke's father? That was some epic cognitive dissonance for young Skywalker. We suffer a similar shock when a new truth is hurled at our beliefs like a wrecking ball.

Do not underestimate the power of cognitive dissonance. It has been responsible for some of the biggest and most embarrassing chapters in our history. To illustrate how pathetic cognitive dissonance makes us look, consider what happened to Galileo. With his telescope in the 1600s, Galileo confirmed the heretical idea that Earth is not at the center of the universe. Such a revelation created cognitive dissonance of biblical proportions because it went against the church's teachings. Rather than accept unequivocal facts, the church chose to cover its eyes and ears. Galileo was sentenced to house arrest and forced to renounce his findings after a meet and greet with the medieval torture devices of his day. The church issued a pardon for Galileo 350 years later, finally ending one of the longest and most shameful cases of cognitive dissonance and denial in human history.

Cognitive dissonance still looms large today because of our personal attachment to beliefs. The best way to avoid being a disgrace or laughingstock to future generations is to detach ourselves from beliefs, even the religious ones we hold so dear. It is important to recognize that they were supernatural theories conjured long ago to soothe a restless brain, but towers of evidence now stand against them. To avoid these trappings, live a life of hypotheses and keep your ideas flexible. Train your brain to get over itself and embrace uncertainty, as these are the necessary steps toward learning. If we live our lives based on evidence available to us at the time, no one can fault that logic. To live our lives ignoring evidence is illogical, and we deserve to be admonished for it.

Dude, Where's My Soul?

The idea that we possess an immaterial soul that outlasts our body is an ancient one. Hundreds of years before the birth of Christianity, Plato wrote about the soul as a nonphysical entity that gives rise to things we can't see, like thought, feelings, memories, and imagination. The concept that we are composed of both material stuff (the body) and nonmaterial stuff (the mind) is dualism, an idea that has resonated with the masses ever since philosopher René Descartes codified it the in the 1600s. Dualism held that the body is divisible and subject to decay whereas the soul is indivisible and lasts forever.

If only Descartes had lived to see what happens in people who have split-brain syndrome, in which their right and left hemispheres are severed to control epileptic seizures. In the 1960s, Roger Sperry's studies of split-brain patients proved that Descartes was wrong. In normal people, an object shown to one side of the brain is seen by both hemispheres. However, because the right and left sides of the brain are no longer connected in these people, objects shown to one hemisphere are not seen by the other. Sperry's work showed that the brain is divisible, like any other part of the body. Even more surprising, each hemisphere operates independently if divided, as if these people have two conscious minds. If we possessed an indivisible soul, this should not happen. Exemplifying that two minds can exist in one brain are split-brain patients whose right and left hands want to do separate things. There are reports of some of these people trying to get dressed in which one hand picks out one outfit and the other hand selects a different one. Renowned neuroscientist V. S. Ramachandran once described a patient whose right hemisphere believed in God but his left hemisphere did not, which would create quite a conundrum for the admissions staff at heaven's gate.

More recently, brain imaging has shown that our "invisible" components—thought processes, memories, and emotions—are actually visible in the brain. It's fascinating to look at a brain scan of someone in real time while they're solving a word problem or having a conversation or, oh yes, having sex. (If the MRI is a-rockin', don't come a-knockin'.) Different areas of the brain light up like an old-school arcade as subjects engage in these various activities. A brain during an orgasm looks very much like a brain on heroin, a finding that informs nonusers just how pleasurable a drug high can be and why it can be so hard for people with addictions to quit.

Like they tested out-of-body experiences, neuroscientists can also evoke potent emotions, sensations, and memories simply by touching different areas of brain tissue with an electrical impulse. By eliciting these responses with an electrical current, we reveal that the brain takes care of all the things a soul is supposed to do. Our intangible thoughts and feelings are indeed made of earthly stuff after all—matter, not magic. In the words of the French physiologist Pierre Cabanis, "The brain secretes thought as the liver secretes bile."

Although these observations constitute formal proof that dualism is wrong, evidence that the mind and body are one has been presented time and time again. Damage to the brain can damage one's personality. The damage can come in the form of a concussion, neurodegeneration, stroke, cancer, or infection. Consider the behavioral changes seen in people with chronic traumatic encephalopathy (CTE), brain tumors, Alzheimer's disease, or rabies. If our soul is immutable and immaterial, physical damage to our brain should not change who we are. If our soul contains our memories and experiences, then the amyloid plaques that form in the brains of people with Alzheimer's should not rob us of them. If our soul was separate from our brain, lobotomies should not work. Anesthesia should not work. Novocain should not work. Perhaps the most damning evidence of all, people can lose or gain religious beliefs due to changes in their brain brought on by neurological disorders or disease.

Even if we get through life with our brain intact, practical questions emerge when we contemplate the idea of an eternal soul. We change as we age, sometimes dramatically. So which version of our soul lives on? Young you or old you? Who does "Jailhouse Rock" in heaven—young Elvis or old Elvis? What about people with severe mental disabilities? Would they still have the disability in the afterlife? If not, wouldn't that fundamentally change who they are? Will your afterlife wife (an afterwife?) be the one who used to laugh at all your jokes or the one that rolls her eyes and tells you to go clean the garage? If your spouse died and you remarried, who is the lucky one that spends eternity with you?

Your essence cannot be captured because it is constantly in flux. You are a Rolodex of personalities—a child, wife, mother, sister, best friend, boss, aunt, Scout mom, soccer mom, cork master at the wine club, tennis player, someone's worst nightmare, and a closet fan of 1980s hair bands. Wearing all these different hats makes you wonder if there really is such a thing as a static, unchanging self. The real Slim Shady can't stand up because we act differently around different people and in different circumstances. Which one of these many versions of you lives on eternally?

If our essence is separate from our body, chemicals that affect the body should not affect our essence. Yet certain mushrooms, LSD, and my mother's meatloaf all produce life-changing hallucinations. Acetaminophen has been shown to decrease empathy. Drugs for Parkinson's disease can turn people into compulsive gamblers. Statins can cause significant mood changes. Nutrient deficiency, dehydration, and fatigue can also have a dramatic impact on how we think and act. If the soul is immaterial, it should be immune to physical substances that alter the body. And yet these substances change our behavior and personality.

As we've learned throughout this book, all of the things that define us, including our thoughts, emotions, and memories, are generated by the brain. Although we may not yet fully understand how these activities function, we know their function does not require a soul. Francis Crick, one of the co-discoverers of DNA in the 1950s, articulated this idea in his 1994 book *The Astonishing Hypothesis*. In Crick's words, "You, your joys and your sorrows, your memories and your ambitions, your sense of personal identity and free will, are in fact no more than the behavior of a vast assembly of nerve cells and their associated molecules." Another notable scientist, Stephen Hawking, has weighed in on the prospect of the soul and afterlife: "I regard the brain as a computer which will stop working when its components fail. There is no heaven or afterlife for broken down computers; that is a fairy story for people afraid of the dark." We all wonder what it feels like after we die, but the truth is we already know. It will be just like the time before we were born. We will have no feeling because we will no longer exist.

That's a splash of cold water in the face! Realizing that we don't have a soul may sound shocking and depressing at first. But living in the light of the truth is better than remaining in the dark. Moreover, good news rises from the ashes of the soul. Both deviant and admirable actions have long been misattributed to an evil or benevolent soul, which takes the focus off research aimed to understand the biological basis for behavior. Undesirable behaviors like violence, addiction, or depression do not stem from immaterial souls; they stem from a material problem with the brain. This is positive news, because the latter we have hope of fixing; the former, we do not.

For soul enthusiasts, the trepidation and anxiety you feel is real. I know; I've been there. Our brain is wary of a world that may be meaningless and governed by chance. We prefer to have faith in baseless ideas like a master plan, karma, or heaven and hell rather than accept that there is no rhyme or reason to what befalls us.

Our brain's misconception that the world is fair is called the "just-world" fallacy: That is, when a random atrocity hurts someone, our brain concludes that the victim must have done something to deserve that fate. If you think a homeless person should just get a job, a person with obesity should just put down the fork, a person with alcoholism should just say no, a woman who dressed scantily was just asking to be raped, or an impoverished nation should just get their act together, you are guilty of the just-world fallacy. You are ignoring contingencies outside of the victim's control that may explain their predicament or inability to do something about it. Blaming a victim not only makes a bad situation worse, but it also distracts us from doing something productive for suffering in the here and now, as well as putting corrections in place to prevent the same from happening to another.

By the same token, our brain employs the just-world fallacy to pat itself on the back for its victories, ignoring the strokes of luck that might have helped us in life. The just-world fallacy explains why some people are apathetic to income inequality. Our brain tends to assume that wealthy people have done something to earn their lot in life and anyone can achieve the lifestyles of the rich and the famous if they just buckle down and apply themselves. Breaking the just-world mentality puts the burden of fixing the ills and inequalities of our world where it belongs: on us.

Realizing that life is a one-take movie without a sequel not only fosters an urgency to live better days—as implied when Steve Jobs said, "Death is very likely the single best invention of life"—but it also puts our petty differences in a new light. I can tell you from experience that your life will not descend into chaos if you abandon the notion of the soul and wrestle yourself free from the grip of all things supernatural. On the contrary, recall from Chapter 5, societies that have freed themselves from religious straitjackets are some of the happiest and healthiest on the planet. In 2005, paleontologist Gregory Paul published a striking analysis showing that secular democracies have lower rates of societal dysfunction compared to pro-religious societies like America.

As articulated by Julien Musolino in *The Soul Fallacy,* the notion of the soul has polluted our ability to devise rational and humane laws governing criminal behavior, addiction, abortion, and the right to die. Concerning human nature, the soul was an incorrect hypothesis. It is one of those ideas that now needs to be retired, and with its disposal comes the promise of a much better understanding of our behavior through the sciences. The dismissal of the soul does not rob us of life's meaning; rather, it leads us to it. As Steven Pinker notes, "Nothing gives life more purpose than the realization that every moment of consciousness is a precious and fragile gift."

Like the Ghostbusters, scientists have eliminated the possibility that there is a ghost in our machine. To view ourselves as something separate from everything else in the cosmos is woefully misguided and unhealthy. Science has shown that we are intimately woven into the fabric of the universe, interconnected with everything and everybody. We are built from genes, but how those genes are expressed depends on our current surroundings and the experiences of our ancestors. Our survival machine is affected by the intimate relationship we have with countless microbes that inhabit our body and cultural memes that inhabit our brain. At conception, our genes and brains could have been shaped in countless ways, but our unique environment and experiences sculpted who we are from that clay. It bears repeating: We did not get to choose our genes or how they were epigenetically programmed. We did not get to choose our microbes. We did not get to choose our brain. We did not get to choose our prenatal or childhood environment, including the belief systems we were taught. So much of what has made us who we are was completely out of our hands. If that does not create humility for oneself and compassion for others, I don't know what will.

What Should We Believe?

We would be better off if we stop clinging to our beliefs, which are like anchors that hold us down. As written by Sent-ts'an circa A.D. 600, "If you want the truth to stand clear before you, never be for or against. The struggle between 'for' and 'against' is the mind's worst disease." In other words, we should not believe in anything; instead, we should draw conclusions based on the available evidence. Our brain gets married to beliefs, and divorcing them is painful. Getting into bed with a conclusion is a more casual affair. The beauty of drawing conclusions is that we can replace them if new data emerge. The diva saves face.

There is no greater waste of time and energy than quibbling over the supernatural. We have more than enough real problems in the real world that urgently need solving. Religions build invisible walls that divide us, creating artificial differences between people. It should not be us against us, but us against the cold, uncaring universe. Perhaps the greatest gift we can leave future generations is to stop filling their fertile minds with spirits and hobgoblins—let's leave our children with a greener mind as well as a greener planet.

» CHAPTER TEN «

MEET YOUR FUTURE

The more we understand our nature,
the better we'll be at nurturing.

—Steven Johnson,
"Sociobiology and You," *The Nation*

A panic gripped New England in the latter half of the 19th century, stemming from harrowing events taking place in Exeter, Rhode Island. A woman named Mercy Brown was proclaimed to be a vampire. Before she died, this mysterious and pale woman roamed about at night, often seen with blood in her mouth or on her clothing. Whoever dared go near her, it was assumed, would soon become a bloodthirsty creature of the night, too.

After she died, fears of vampirism continued to spread like wildfire through the town. Rumor had it that Brown was returning from the grave at night to feed off the living, including her younger brother. The frenzied townspeople exhumed her body shortly after its burial and found coagulated blood in the corpse's heart. Believing that this proved she was a vampire, they ripped out her heart, burned it, and fed the ashes to her ailing younger brother. Alas, their "treatment" did not work and the poor boy died shortly thereafter.

Meanwhile, in Germany, a scientist named Robert Koch was working to identify the cause of a strange condition called consumption. Consumption, which we now call tuberculosis, is a contagious and progressive lung disease that causes pale skin, insomnia, and the coughing up of blood. Some of his peers mistook the symptoms for vampirism, but Koch dismissed such nonsense in favor of a down-to-earth explanation. In 1882, his hard work revealed the true cause of this condition: a bacterium he named *Mycobacterium tuberculosis*.

As we've learned, science has trounced the traditional view that our behavior stems from a nebulous spirit dwelling within the body. The revelation that our actions have a mechanical, biological basis produces mixed emotions among people accustomed to thinking that there must be something more to the human equation than cells and biochemicals. When it comes to explaining our behavior, some of us are kindred spirits of Robert Koch, convinced that our actions are products of our biology. Others remain shackled to the mentality of the Exeter villagers, failing to realize that supernatural explanations are intellectual quicksand.

Throughout this book, we have uncovered many cryptic ways that genes, epigenetics, microbes, and our subconscious influence our personality, beliefs, and virtually everything we say and do. Now that we have thrust these hidden forces into the light, can we devise ways to outwit them? The revelations have been surprising. But the good news is that knowing the truth underlying our behavior is a necessary prerequisite to doing something about it.

Wouldn't it be great to modify genes or microbiota that predispose people to obesity or substance abuse? Can we pop a smart pill or implant tech into our brains to elevate our cognitive abilities? How about using this knowledge to cure mood disorders or criminal behavior? These sound like lofty goals, but they might be more within our reach than we realize.

How We Can Change Our Genes

The power to alter our genetic constitution has the potential to address a wide range of issues, from the trivial (helping supertasters enjoy broccoli) to serious (fixing the gene variant that causes Huntington's disease), to downright controversial (adding genes that augment intelligence or put eyes in the back of our head).

Technically, we've been making genetically modified organisms (GMOs) for more than 10,000 years by selectively breeding plants and animals. Through selective breeding, we've taken the wheel of evolution and steered life into forms that better suit our purposes. Just to name a few, we've made bigger tomatoes, sweeter apples, tamer dogs, and plumper chickens. The process is agonizingly slow and, as our lack of pet bears attests, not all species are amenable.

Since the discovery that DNA is the recipe for life in the 1950s, scientists have been cooking up more efficient ways to improvise on this recipe. They started small—really small—by altering bacterial cells called *E. coli*. In 1973, Stanley Cohen and Herbert Boyer were able to make bacteria absorb and read foreign DNA. They created the first genetically engineered life-form by inserting a piece of frog DNA into *E. coli*. Not a very practical creation, but it was an important leap. The bacteria read the frog DNA and made a frog protein from it. We then started employing bacteria to make proteins we can use, including insulin, human growth hormone, and proteins for vaccines. One year later, Rudolf Jaenisch and Beatrice Mintz made the first genetically modified animal by injecting a gene into mouse embryos. Since then, scientists have genetically modified plants, fungi, roundworms, fish, insects, rats, monkeys, and more. Next stop: people.

Ironically, our initial attempts to modify DNA in people, referred to as gene therapy, coincided with the release of the film *Gattaca* in 1997. Gene therapy involves giving a person a working copy of a bum gene. Although that sounds as easy as a Matchbox Twenty guitar solo, it has proven to be a real challenge. You just can't swallow a gene pill, because the gene won't work unless it's inside the cells that need fixing. If only we could shrink down doctors to a microscopic level so they could travel through a patient's body in a tiny submarine, like in the movie *Fantastic Voyage,* perhaps they could deliver the gene only into the cells that need it. It seemed impossible, but then scientists had a eureka moment: Viruses behave just like that miniature submarine, taking their DNA cargo only into the cells they infect. Perhaps we can pimp viruses to deposit therapeutic genes into a patient's cells.

Domesticating viruses is like dancing with wolves; they can deliver genes into us, but they remain an unpredictable infectious agent that can bite. The field ground to a halt in 1999, when 18-year-old Jesse Gelsinger died in a gene therapy trial. The researchers injected him with adenovirus carrying a good copy of the gene Gelsinger needed, but the adenovirus induced a lethal immune response days later.

In 2000, gene therapy cured several children born with a tragic immune deficiency so severe that they were not able to leave the confines of a sterile or germ-free room (a condition commonly known as "bubble boy" disease.) However, the treatment also caused a leukemia-like disease in some of the recipients. In this trial, a retrovirus was used because it not only delivers the good gene into cells, but it also sews the gene into the fabric of the patient's DNA, making it a permanent resident. Unfortunately, in the process of stitching the gene the patients needed into their DNA, another gene was damaged that put them at risk of cancer.

This frustrating combination of success and setbacks has called the use of viruses into question. Can we really tame these tiny wild beasts? Over the past two decades, researchers have discovered new ways to disarm viruses, like Hermey the Elf pulling the teeth out of Bumble the Abominable Snow Monster. This hard work has paid off, and gene therapy is making a comeback.

In 2017, gene therapy obtained a resounding victory with the treatment of adrenoleukodystrophy or ALD, a condition featured in the movie *Lorenzo's Oil.* ALD is a rare and incurable neurodegenerative disease that strikes otherwise healthy children around the age of seven. As the disease progresses, children lose the ability to control their muscles, making it impossible to walk, talk, or eat without a feeding tube. ALD is caused by a mutation in a gene called ABCD1, which makes a protein that sends fat molecules to their degradation chamber in brain cells; when this protein isn't working, these fats build up and induce an inflammatory response that damages the brain.

The concept to fix ALD is simple: Give patients a normal copy of the ABCD1 gene so these fats can be broken down as nature intended. Researchers successfully adapted a retrovirus called lentivirus to deliver a working copy of ABCD1 into bone marrow stem cells removed from the patient. After the gene is inserted into these stem cells, they are infused back into the patient. Stem cells are undifferentiated cells, which means they have the potential to become any cell type in the body. Some of these genetically engineered stem cells developed into brain cells now capable of disposing the problem fats.

Another victory for gene therapy quickly followed with the successful treatment of epidermolysis bullosa (EB) in a seven-year-old child. EB is a rare condition in which the skin is very weak and prone to tears that easily become infected. In almost half the cases, kids with EB don't live long enough to get a driver's license. Scientists used gene therapy to correct the mutant gene in stem cells taken from the patient. The repaired stem cells were developed into skin cells and grown in the laboratory until they were large enough sheets to be grafted onto the patient.

With the development of safer and more efficient viral delivery systems, gene therapy has shown promising results in beta-thalassemia, some forms of inherited blindness, the bleeding disorder hemophilia, and more. And more success stories are on the horizon as we expand our gene editing toolbox. A state-of-the-art gene editing technique called CRISPR/Cas9 has generated so much excitement that it has nearly become a household term. Derived from components of a bacterial immune system, CRISPR/Cas9 operates like DNA scissors that cut with nucleotide-level precision to destroy bad genes or insert new genes.

In 2015, Junjiu Huang and his team at Sun Yat-sen University in China created the first genetically modified human embryo using CRISPR/Cas9 to fix a bad copy of the beta-globin gene that causes beta-thalassemia. (The embryos used in this study were not viable.) Analogous to CRISPR/Cas9, zinc finger nucleases (ZFNs) can also cut specific sites of DNA to insert new genes. Another gene therapy tactic under intensive investigation is immunotherapy, which involves the genetic engineering of a person's immune system so it is able to recognize and battle cancers such as lymphoma. One immunotherapeutic approach called CAR (chimeric antigen receptor) T cell therapy involves the removal of a patient's T cells, which are then edited at the genetic level to make a specific receptor used to recognize that patient's malignant cells. These modified or "reprogrammed" T cells then act liked hired assassins when infused back into the patient, hunting down and killing cancer cells.

The advent of gene editing tools means that we are no longer just readers of DNA; we are scribes (although presently, we're still learning the language). For the same reason we don't ask a preschooler to edit our manuscripts, most scientists advocate banning gene modification in viable embryos or sex cells, because such modifications would be heritable and could have unpredictable adverse effects on the individual and potentially future generations. In addition, it opens up an epic can of ethical worms because it won't be long before someone wants to edit genes that go beyond medical benefits, creating designer babies.

How We Can Change Gene Expression

We have discussed many behavioral issues that do not arise from variation in gene sequences, but rather from a difference in the amount of gene expression. The environment can influence gene expression levels through epigenetic mechanisms such as DNA methylation or the chemical modification of the histone proteins that interact with genes. As scientists uncover the enzymes that write, read, and erase these epigenetic modifications, it has become apparent that we can target them with drugs. The basic premise: Gene X is getting methylated and shut off. Bad things happen when gene X is turned off. Let's turn gene X back on with a drug that prevents DNA methylation.

Although epigenetics is a new field of study, the Food and Drug Administration (FDA) has already approved several epigenetic drugs to treat various illnesses. The first, in 2004, was azacitidine, which treats myelodysplastic syndrome (MDS), a rare bone marrow disorder that received widespread attention when *Good Morning America* co-anchor Robin Roberts announced her diagnosis. Azacitidine inhibits a DNA methylation enzyme, and the resulting decrease in DNA methylation amps up gene expression. Although not selective in which genes are cranked up, the subset of genes needed for blood cells to mature is affected, thereby raising blood cell counts and alleviating symptoms in some patients.

In 2006, the FDA approved a second class of epigenetic drugs called histone deacetylase (HDAC) inhibitors. These drugs are used to treat lymphoma or multiple myeloma, but newer derivatives to treat solid tumors are in clinical trials. Recall that the genes associated with acetylated histones are actively expressed. When HDAC enzymes remove acetyl groups, the gene's expression is reduced or shut off. HDAC inhibitors stop the enzymes that remove acetyl groups from histones, thereby keeping the gene in its active state.

How might HDAC inhibitors work against cancer? Our DNA is equipped with tumor suppressor genes, which make sniperlike proteins that detect and kill treasonous cells. Aggressive cancer cells avoid this fate by shutting off tumor suppressor genes, and it has been proposed that HDAC inhibitors may keep these tumor-fighting genes on. Because changes in gene expression are associated with many types of diseases, there is hope that HDAC inhibitors may prove useful against neurological disorders like schizophrenia, metabolic disorders like obesity, cardiovascular disease, or even reversal of aging.

Epigenetic drugs may be able to alter the way DNA is programmed prenatally or during early childhood. Recall the experiments of Michael Meaney from Chapter 5 that showed pups born to neglectful rat mothers have higher levels of DNA methylation, which deactivated genes needed for proper stress responses and caused the pups to be overly anxious. By giving the pups HDAC inhibitors, Meaney was able to reverse these behavioral problems by turning up the volume on the stress response genes to where it should be.

In the comic strip Calvin and Hobbes, Calvin used his cardboard box "transmogrifier" to turn himself into an elephant so he'd remember his vocabulary words. Surely there must be a less obtrusive way to boost our memory skills. It turns out that epigenetic drugs may transmogrify our genes in ways that enhance learning and memory. In rodent models of cerebral ischemia (stroke) and Alzheimer's disease, administration of HDAC inhibitors minimizes brain damage and enhances memory retrieval and retention. HDAC inhibitors also boost memory and learning in normal rats that are not suffering deficits due to disease.

A 2015 study by neuroscientist Kasia Bieszczad at Rutgers University showed that an HDAC inhibitor fostered stronger neuronal connections, which may explain the enhanced memory observed in rats receiving the drug. The idea is that when we are focused on a task, a gene network in our brain that forges memories is activated, in part through the acetylation of histones associated with those genes. HDAC inhibitors stop the enzymes that remove these acetyl groups, which means the gene network needed to form memories stays active longer.

As with all epigenetic drugs, specificity is a problem. They may be like the bratty kid in an elevator who presses every single button instead of just the floor he needs; in other words, epigenetic drugs may affect all genes and not just those that need modification. One of the pioneers in the field, neuroscientist Li-Huei Tsai at MIT's Picower Institute for Learning and Memory, is well aware of this problem and is working to identify precisely which HDAC enzyme(s) from among the 20 or so in our body is involved in memory. In 2009, she and her team found that HDAC2 is a negative regulator of memory formation in mice, which suggests that an HDAC inhibitor that only targets HDAC2 may have fewer side effects.

Another way to take control of our gene expression that does not rely on drugs is changing our environment, which includes diet and exercise. As we've seen, our environment can have important influences on which genes are turned up or down. By modifying your lifestyle, you can exert a level of control over your gene expression. Exercise is the best medicine to ward off many health ailments. We know that exercise strengthens muscles, protects the heart, keeps cholesterol in check, and helps maintain a healthy weight. What we have not appreciated until recently is that exercise also changes gene expression through epigenetics.

While you are huffing and puffing in the gym, that physical exertion is putting your epigenetic machinery to work to reprogram your genome. Scientists at the Karolinska Institute in Stockholm had participants exercise with only one leg for 45 minutes four times a week for three months—the other leg got to be lazy. I imagine the participants were easy to spot, hobbling down the street with one meaty leg and one scrawny leg. The study revealed thousands of differences in DNA methylation between the exercised leg after the training period compared to before the training, including differences on genes associated with healthier metabolism and immune responses. The slacker leg showed no significant differences in DNA methylation before or after the training period.

Numerous studies have shown that physical activity doesn't just beef up our muscles, but it also beefs up our brain. One way that exercise can benefit our noggin is through the epigenetic changes it makes. Remember the previously mentioned HDAC2, the histone deacetylase that has a negative effect on learning and memory? A 2016 study found that exercise produces a biochemical in the body called β-hydroxybutyrate, which happens to be an HDAC inhibitor that targets HDAC2. Consequently, the inhibition of HDAC2 promotes expression of BDNF (brain-derived neurotrophic factor), a protein known to enhance memory and stimulate growth of neurons.

The epigenetic benefits of exercise may not be limited to improving your brain but may also help your kids become smarter. A 2018 study led by geneticist André Fischer at the German Center for Neurodegenerative Diseases in Göttingen, Germany, showed that male mice that exercise produce sperm that are epigenetically different than sperm from lazy mice. Male mice were divided into two groups: One group was placed in an empty cage while the other was placed in a cage set up like a mouse gym. The physically fit mice sired higher achievers with greater learning ability than fathers that did not exercise. These smarter offspring showed improved communication between neurons in the hippocampus, a region of the brain important for learning. It is believed that the epigenetic differences seen in sperm from exercising fathers benefitted brain development in the pups.

In addition to exercise, you may want to consider mindful meditation, which is a bit more involved than chanting "Serenity now!" as Frank Costanza from *Seinfeld* was advised to do when he experienced stress. Mindful meditation involves being quiet and still while focusing on nothing but your breathing—the type of meditation Buddhist monks and Jedi Knights practice. Researchers have found that mindful meditation is associated with a reduction in HDAC2, altered histone acetylation levels, and decreased expression of pro-inflammatory genes. These findings are beginning to reveal the biological basis for why meditators often handle stress better than others.

How We Can Take Control of Our Microbial Guest List

We've seen numerous instances in which the bacteria, fungi, and parasites residing in our body affect behavior in startling ways. These microbes produce thousands of biochemicals, including neurotransmitters that can affect how we feel, what we crave, and how we act. As we continue to learn more about which microbe does what, thoughts have already turned to how we might be able to manipulate these microbes to our advantage. After all, as the host of these organisms, shouldn't we have a say in who we invite to our shindig?

Some of our microbes are clearly unwanted guests, like the *Toxoplasma* parasite that loiters in the brains of about a third of the population. Unlike the friendly bacteria that share a symbiotic existence with us, pathogens like *Toxoplasma* do not belong and should be evicted. The problem is this parasite resides *inside* of our cells, surrounded by a fortress of parasite proteins called a cyst wall. To kill this parasite, a drug first has to gain access to the brain, which is shielded by a blood-brain barrier that shouts, like Gandalf, "You shall not pass!" Second, the drug has to get inside the infected neurons. Third, the drug has to get through the thick cyst wall encasing the parasites. Finally, the drug has to get into the parasites themselves. This is a lot to ask of a drug.

Nevertheless, my laboratory at the Indiana University School of Medicine has been investigating experimental treatments aimed at eliminating *Toxoplasma* brain cysts in mouse models of infection. In 2015, studies led by Imaan Benmerzouga discovered that guanabenz, an old blood pressure medication that crosses the blood-brain barrier, is able to significantly reduce the number of parasite cysts in the brains of infected mice. It is hoped that drugs like guanabenz may do the same in human patients. But for now the best way to manage *Toxoplasma* is not to get infected in the first place, which involves responsible cat ownership, feral cat management, and proper preparation of food products.

As for the symbiotic microbes inhabiting our guts, researchers are currently working to identify which species produce which effects on our well-being and behavior. Pioneering studies to modify our microbiota with probiotics have produced some promising results. In addition, "fecal transplants" have emerged as a means to get desirable bacteria into someone's innards. The most successful application to date has been in the treatment of *Clostridium difficile,* or *C. diff.* This bacterium, which is naturally resistant to most antibiotics, can grow out of control and wreak havoc on the colon when benign bacterial species are depleted (which can happen, for example, when patients take antibiotics for prolonged periods). Restocking the colon with bacteria from a healthy donor reestablishes a gut microbiota that can keep *C. diff* in check. The success of this treatment has spurred the question: Can we manipulate our gut microbiota to achieve other health benefits?

Neuropharmacologist and microbiome expert John Cryan at University College Cork in Ireland thinks so. In fact, he believes doctors will soon test our microbiota along with routine blood tests during checkups. Cryan also envisages design and usage of bacteria-based drugs called "psychobiotics" comprised of live microbes suspected to have a positive effect on mental health. Psychobiotics would be like ordering a fecal transplant, but hold the feces. The medicinal bacteria and fungi can simply be grown as laboratory cultures and then processed like other probiotics, such as those taken for digestive health. The key questions are which microbes to include, how many, and will they really work on the brain without producing side effects?

At this stage, scientists are laboring to find connections between certain species of bacteria and their effect on the brain and behavior. A combination of two bacterial species, *Lactobacillus helveticus* and *Bifidobacterium longum,* given in the form of a probiotic, has been shown to reduce anxiety in people by decreasing the stress hormone cortisol. Other species of bacteria like *Bifidobacterium infantis* have antidepressant properties in rats. A great deal of attention is also being focused on whether gut bacteria are linked to symptoms associated with autism spectrum disorders. A provocative 2013 study by biologist Sarkis Mazmanian at California Institute of Technology showed that administration of bacteria called *Bacteroides fragilis* reversed autistic symptoms in a mouse model.

The status of our microbiota may also impact our susceptibility to trauma, whether experienced in childhood or on the battlefield. Physiologist Christopher Lowry at University of Colorado performed a study in 2017 that compared the species of microbiota within people suffering from post-traumatic stress disorder (PTSD) with people who experienced similar trauma yet did not develop PTSD. Individuals who experienced childhood trauma and adults with PTSD are deficient in several species of bacteria, including Actinobacteria, Lentisphaerae, and Verrucomicrobia. These bacteria work to balance the immune system, and their loss may partly explain why individuals with PTSD often have problems like inflammation.

There may be other ways to promote the growth of helpful bacteria in our guts, such as prebiotics. Prebiotics are components of our diet, like fiber, that help grow a favorable population of microbes in our gut. Much like fertilizer supports a healthy garden, what you eat supports a healthy microbiota. Most prebiotic regimens follow commonsense guidelines for healthy eating: including a bevy of fruits and vegetables and avoiding processed foods loaded with sugar, salt, and fat.

The potential for prebiotics was illustrated in a 2017 study that tested whether 12 weeks of eating a Mediterranean-style diet would help people suffering from major depression. Following this diet helped improve symptoms in 30 percent of participants. Many studies are currently under way to determine how the prebiotics in healthy foods modulate our microbiota to produce these beneficial effects on mood. Although dietary changes alone are not likely to be completely effective in most people with depression, prebiotics and probiotics (the combination of which is called synbiotics—as if there are not enough biotics already) may soon be an important component to psychiatric treatment.

Taking control of our microbial guest list includes inviting bacteria that bring something we need to the party. Scientists are creating bacteria that may help people suffering from phenylketonuria (PKU), a rare genetic disorder that makes eating protein nearly impossible because these sufferers lack the enzyme needed to break down the amino acid phenylalanine. Synlogic is a company that engineers bacteria to carry the gene encoding the enzyme PKU. The idea is that people with PKU may be able to eat protein if they also consume this phenylalanine-busting bacteria.

It is important to note that the excitement surrounding genetic engineering, epigenetics, and the microbiota makes these cutting-edge areas of science ripe for overhype. These fields are in their infancy, and much more research is required to substantiate the promises of early studies. At this time, not enough evidence exists to support the claims that some shyster's snake oil supplement, alternative medicine, or strange New Age activity is going to modulate your genome, epigenome, or microbiome in a healthy way. On the contrary, it could do harm.

How We Can Hack Our Brain

The revolution to combine brains and electronics is already under way, triggered in the 1960s by José Delgado, the neuroscientist who stopped a charging bull simply by pressing a button on a remote control (Chapter 6). Thirty years later in the 1990s, Phil Kennedy spearheaded the first effort to merge a computer with the human brain.

The subject was named Johnny Ray, who was completely paralyzed by a stroke in his brain stem at age 52. Whether caused by stroke, amyotrophic lateral sclerosis (ALS), or a tragic accident, patients like Ray are referred to as "locked in" because they are fully conscious within a body they can no longer move. By connecting electrodes between Ray's brain and a computer, Kennedy created the first brain-computer interface (BCI). Seemingly like magic, Ray was able to use the BCI to move a computer cursor across the screen using only his thoughts. Shortly thereafter, Ray's thoughts could move the cursor to letters on the screen to type words, allowing him to converse with loved ones for the first time since his stroke.

Johnny Ray bears the distinction of being the first "human cyborg," a term referring to the meshing of biological machines and manufactured machines. And, unlike the nightmarish visions depicted in science fiction stories, the fusion of computers with our mind did not rob Ray of his humanity, but rather helped to restore it.

In 2006, neuroscientist John Donoghue at Brown University invented BrainGate, a tiny implant containing a hundred electrodes embedded into the motor cortex of a patient with quadriplegia. The BrainGate array could then be connected to a computer through a portal sticking out of the top of his head. When connected to the computer, he could use his thoughts to open email, play Pong, and change the channels on his television.

Scientists at the Applied Physics Lab at Johns Hopkins University are using BCI to develop modular prosthetic limbs. In patients missing a limb, signals from their brain are sent to a custom-built socket trained to move a prosthetic accordingly. The latest robotic arm has no less than 26 joints and can curl up to 45 pounds by thought control. Current studies are focused on sending information in reverse, from limb to brain, so that amputees can feel sensations, such as the texture of Velcro or temperature of soup. Similarly, other groups are trying to help people who are blind by installing brain implants that can interpret visual information brought in from a camera headset.

Thank goodness, scientists are developing BCI technologies that avoid invasive brain implants in favor of electrode grids laid on top of the brain's surface (electrocorticography or ECoG)—or, even better, on the scalp (electroencephalography or EEG). EEG electrodes placed on the scalp resemble a swim cap with more than a hundred wires attached that feed into a computer. Both scan a wider range of brain activity and attempt to translate these patterns into speech or motion, very much like mind reading.

This feat becomes possible because each one of the billions of neurons comprising our brain emits a tiny electrical pulse when activated. When groups of neurons are activated, they collectively emit neural oscillations, otherwise known as brain waves. Distinct patterns of brain waves emerge depending on what the person is thinking, and an EEG can read and translate the diagram of electrical activity into action by a computer. Imagine flying a drone just by thinking about it: This amazing technology has turned such a fantasy into reality.

BCIs have also been used to control other people through thought, even if they are in two different locations. Researchers were able to make one person use his thoughts to control how another person played a video game that involved shooting targets. Person one wore an EEG cap and *imagined* pressing the fire button when the target appeared during the game. The brain waves his thoughts generated were sent over the Internet to the second person, who was wired up to the computer but with his back to the video game. Even though he was in a different building and couldn't see the video game, the second person was able to shoot the targets accurately because of person one's thoughts. By the time the *Avatar* sequels finally come out, they might not be science fiction anymore.

Phil Kennedy speculates that one day we will be able to incorporate the brain into a robotic body, providing the means for us to live forever, free of flesh. How he intends to transfer the influence of microbiota on our mind is not clear. But you might want to start stocking up on Rust-Oleum instead of skin cream.

Another branch of neuromedicine that sprung from Delgado's classic experiment in the bullring is deep brain stimulation (DBS). DBS is now routinely used to treat patients with major depression, obsessive-compulsive disorder, or movement disorders like Parkinson's disease. Sometimes compared to pacemakers for the heart, DBS involves implanting electrodes that emit electrical pulses into the brain. Although the precise mechanism is still under investigation, it is thought that the electrical impulses from the device disrupt or reset the problematic electrical activity occurring in the brain that leads to the patient's condition.

The placement of the electrode in the brain is dictated by the neurological disorder being treated. The electrode is a long, thin needlelike probe that is inserted into the brain while the patient is awake (unless the patient's motion disorder prevents her from staying still). As disconcerting as that sounds, this is not a physically painful procedure, because the brain has no pain receptors. Additionally, having the patient awake is helpful because she can verbally let the surgeon know if the system is working. (Similarly, if the patient proclaims she is eager to get back to Qo'noS to reclaim her seat on the Klingon High Council, the surgeon knows to reposition the electrode.)

As you get older, you may struggle to recall the details of your cherished memories. Is there an implant for that? Biomedical engineer Theodore Berger at the University of Southern California is one scientist working on memory boosters that can interface directly with our brain. As intangible as it may seem, memory is a biological phenomenon conveyed by electrical impulses between our neurons (see Chapter 8). A part of the brain called the hippocampus converts working, short-term memory into long-term memory. In theory, if we learn how to read the language of the brain waves produced by this electrical activity, we should be able to decode a memory. On the flip side, it should also be possible to send memories into the brain through the hippocampus, similar to the process Arnold Schwarzenegger's character experienced in the 1990 movie *Total Recall.* However, the types of memories we would send might include the collective works of Shakespeare, how to speak a foreign language, or a reminder of what happened in the previous season of our favorite show.

Memory problems sometimes arise because of a signaling defect in the hippocampus; this is where Berger and his team have gone to learn the language of memory. They worked in rats and monkeys to record the electrical signals emanating from the hippocampus while the animals learned a simple task, like which lever to push for a treat. These electrical signals were programmed into a memory chip that was implanted into the hippocampus. With the memory chip turned off, the animals were given a drug that blocks long-term memory retrieval, causing them to forget which lever yielded a treat. But when the memory chip was turned on, the drugged animals knew which lever to press for the treat. These promising results are hoped to pave the way to help people with memory problems.

Sometimes the problem is not being able to forget a memory. Upward of 8 percent of Americans suffer from PTSD, an often debilitating disorder caused by unshakeable memories of traumatic experiences. As we learned in Chapter 8, memories are reconstructed every time they are recalled, and these memories can be distorted as they are reconsolidated (written back into our memory bank). Scientists have speculated that it may be possible to exploit this process by altering the memory during its reconsolidation. One way to do this could be through drugs like beta-blockers, which reduce heart rate and anxiety. A 2009 study by psychologist Merel Kindt at the University of Amsterdam administered mild shocks to volunteers at the same time they viewed pictures of spiders. A day later, they gave half the group a beta-blocker pill and the other half a fake pill before reactivating their memory by showing them the spider photos again (without shocks); both groups were equally startled by the pictures. The surprise came when they were shown the spider pictures again a few days later. Those who received the beta-blocker during memory reactivation were not as startled when shown the spiders. In contrast, the volunteers who took the dummy pill remained very jumpy when seeing the spider images.

Not unlike what you might have seen in an episode of *Black Mirror,* scientists are also trying to move the Internet into our heads. A few decades from now, teens will laugh, "How lame that people had to carry, like, a device to post dessert on Snapchat!" Many people already use the Internet as a second brain, quick to ask Google a question, check their calendar, or stroll down memory lane by viewing photos on their Facebook timeline. It would not be much different than fusing your smartphone directly to your brain instead of feeding it to your brain through your eyes. Wouldn't it be great to have these resources available to our thoughts? Or at least get a hard drive connected to our brain for better memory storage and recollection? Scientists predict new technologies that augment memory and intelligence will be online in the next 30 to 50 years. A new breed of cyborg-based viruses and malware should follow shortly thereafter, and a Ph.D. in neuroscience may be a prerequisite for future IT personnel.

Why We Need to Live an Evidence-Based Life

Editing genes, modulating gene expression with epigenetic drugs, managing our microbiota, and brain-computer interfaces are exciting technologies that will improve lives down the road. Some will refer to these marvels of technology as miracles, but that is a mischaracterization, for these advances are a direct result of employing scientific method. They are the reward that comes from centuries of arduous investigative research aimed at understanding the mechanics underlying human physiology. Instead of this reductive knowledge diminishing the majesty of humanity, it is helping us to alleviate suffering and improve lives.

At 117 years of age, Emma Morano was one of the oldest people in the world until she passed away in April 2017. She was born during the New England vampire panic, putting into perspective how recently this episode took place. To paraphrase Carl Sagan, we have not stepped far out of the shadows of our demon-haunted world; in fact, too many people are still languishing in the darkness of those shadows. In October 2017, mobs in Malawi killed eight people believed to be vampires.

The progress made in our battle against infectious disease illustrates the difference between scientific behavior and superstitious behavior, and emphatically underscores which is more beneficial to our future well-being. Not too long ago, supernatural forces were believed to caused life-threatening plagues. An angry god, a spiteful witch, monsters, or a violation of some superstition were usually cited as the reasons why individuals or a village fell ill. Thoughts and prayers, the burning of innocents accused of witchcraft, and blind faith in pointless (and sometimes harmful) superstitions were remedies for our troubles, and they were all woefully unsuccessful.

But the discovery of germs in the mid-1800s revealed the true reason why people coughed up blood or became riddled with nasty pustules. Understanding the biology underlying disease positioned us to do something tangible about the problem, leading in the early 1900s to the discovery of the antibiotic penicillin, heralded as a "miracle drug" at the time. Penicillin has saved untold billions of lives. But to call that a miracle is a slap in the face to the long chain of diligent and curious scientists who labored to uncover the real reasons people got sick. They were the courageous ones who rejected the traditional hypothesis that ascribed illness to the supernatural. When we concede a problem to the realm of the supernatural, nothing helpful can be done. But when we roll up our sleeves, experiment, gather evidence, and think critically, then progress is made.

The same principles hold true for our unexplained behaviors today. We are who we are, and we do the things we do for a logical reason embedded within our biology. Identifying the true sources of our behavior sheds light on who we really are, and what we are capable of becoming. Heeding the data allow us to live a life guided by evidence instead of assumption.

One could cite many examples that illustrate the benefits of living an evidence-based life. But the ones I find most compelling relate to improving the lives of young people. Time and time again in these pages, we've seen how absolutely critical the prenatal and early years of life really are in ensuring happy and healthy adult lives. Let's look at one case where evidence-based living was applied to the problem of unruly teen behavior and substance abuse.

Do you know which country has some of the most well-behaved teenagers? Iceland. But this wasn't always the case. In the 1990s, over 40 percent of Icelandic teens were drinking and nearly 20 percent using pot. Today, those percentages are down near 5 percent. How did Iceland achieve this success? It wasn't religion or zero tolerance policies for substance abuse; it was understanding biology. Remember the famous Rat Park experiment described in Chapter 4? Give the rats plenty of fun and interesting things to do and they avoid the cocaine placed in their environment.

In the 1990s, officials in Iceland attempted to do something similar with Project Self-Discovery, a program that offered teens the chance to experience natural highs instead of drug-induced ones. State-sponsored after-school programs were put into place that gave teens opportunities to learn something new, like playing the piano, sculpturing, or learning how to tango. They could take martial arts or play sports. These are activities that many families simply could not afford until the implementation of this program. In addition, the kids attended life skills training, and parents attended sessions that provided tips on raising teens. A curfew was also implemented to keep teens from being outside after 10 p.m.

In the United States, many well-to-do families have the luxury of doing such things, and it generally pays off; studies show that the teen mind craves dopamine, supplied in sufficient measure through these extracurricular activities. Imagine what would happen to crime and drug addiction rates if all school districts had equitable resources that offered children natural highs. Investments that provide each child with proper nutrition, direction and mentorship, and knowledge about drugs and sex are much cheaper than the cost of dealing with problem adults. It's not just a humane thing to do, but also the most economical.

Similar thoughts crossed the mind of David Olds at the University of Colorado Health Sciences Center, who tested how far a little care and education can go in the rearing of children. Olds enrolled 400 first-time pregnant mothers from lower-class neighborhoods in New York in his study. Participants were treated to in-home visits by health professionals about 10 times during the pregnancy and about 20 times during the first two years of the child's life. The progress of their children was then monitored 13 years later, when they turned 15. During the visits, the mother was counseled about proper nutrition for her and her baby and was also taught parenting skills.

The results exceeded expectations, showing that these simple and inexpensive prenatal and early childhood home visits dramatically reduced the number of subsequent pregnancies, the use of welfare, child abuse and neglect, and criminal behavior. This evidence was published in 1997. Maybe one day we'll become wise and compassionate enough to use it.

Science has dispelled the idea that anyone can be anything they want to be; great disparities in nature and nurture put us on a vastly unequal playing field. But we can take practical measures to minimize these disparities and help everyone live to their full potential. When it comes to our fellow human beings, especially our children, the choice should not be sink or swim; it should be swim or be rescued. Informed by our biology and guided by evidence, we can build better environments for all, which will lead to a stronger and healthier society.

» CONCLUSION «

MEET THE NEW YOU

Instead of condemning people, let's try to understand them. Let's try to figure out why they do what they do. That's a lot more profitable and intriguing than criticism; and it breeds sympathy, tolerance, and kindness.

—Dale Carnegie, *How to Win Friends and Influence People*

As you can see, understanding why we do the things we do is really simple. Not! Nothing explains everything, meaning that there is never a single explanation for our behavior. Nor can our success and failures in life be attributed solely to our awesomeness or lack thereof. Our actions and personalities arise from a dizzying interplay between genes (including how they've been epigenetically programmed), microbes, hormones, neurotransmitters, and our environment. We also cannot view our present behavior without recognizing the dirty fingerprints left behind by the evolutionary pressures that shaped us—especially the intense subconscious drive to survive and reproduce.

But the jig is up. We now know that we're elaborate survival machines for genes, created to continue their billion-years-old game of replication. It hasn't been easy meeting our DNA maker and getting wind of all the tricks it has used to dupe us into keeping genes alive. Prince or pauper, we're all slaves to the DNA grind. We're like Pinocchio waking up to the gift of life, but dismayed to find strings attached.

After all these years of thinking we were free agents, we've come to realize that most, if not all, of our behavior is not of our own volition. It has been guided and restricted by puppet strings. One string is DNA. Another is epigenetics. Another is our microbiota. Yet another is our subconscious. And we're discovering more strings that tug on our behavior in ways still unbeknownst to us. For example, when a gene is transcribed into a protein, the genetic instructions are carried on a molecule called messenger RNA (mRNA), which can also be chemically modified like DNA can be methylated. The study of chemical modifications on mRNA is called epitranscriptomics, and these changes affect how much protein is made from the mRNA and when. Proteins themselves can also be chemically modified in ways that alter their stability, function, or location in a cell. All these additional regulatory steps make it harder and harder to predict someone's behavior solely on their gene sequences.

Once concealed from our view, the puppet strings that have been controlling us are now visible. More than that, we're discovering potential ways to cut the cords through gene editing, epigenetic drugs, remodeling microbiomes, and merging brains with computers. The hand of fate that has served as our puppet master has been evolution. But like a puppet that has learned to be its own puppeteer, science has provided us the ability to evolve. Only time will tell whether humanity's puppet show will be a hit or a flop.

We stand a better chance of success if we allow history to serve as our Jiminy Cricket. Human nature was born from selfish genes, but selfish genes are so yesterday. The disease of selfish genes continues to plague our species in the form of inflated egos, greed, dishonesty, cheating, the creation of "us versus them" dichotomies, and the tolerance of a social hierarchy that allows the world's riches to be restricted to a few alpha males while billions wallow in poverty. These selfish genes created the diva part of the brain: the part that almost always gets us into hot water or causes suffering in others.

But these genes have also created a brain that devised the scientific method as a means to better understand itself and the universe it inhabits. Over the ages, from astronomy to zoology, science has been taking us off the pedestal upon which the diva brain has perched itself. The comic strip Calvin and Hobbes brilliantly captured coming to terms with this reality: Calvin defiantly shouts at the starry night, "I'M SIGNIFICANT!" but then mutters, "Screamed the dust speck."

Yes, science is an ego crusher, but a kick in our complacency is a humbling unifier that does our diva brain good. Egos put up unnecessary walls between the folks we consider to be on our team and the other people in the world. The demolition of the ego will help us erase the nonsensical lines dividing us, turning fists into open hands.

History has proven that cooperation is infinitely more beneficial for individuals and society at large. Species that learn to team up and divide labor are mimicking what individual genes did long ago when they came together as a team in DNA. A little autonomy is sacrificed for the greater good. But because individuals (and their kin) often benefit from that greater good, it forms a positive feedback loop.

The vast majority of nature is red in tooth and claw, with no regard for the welfare of others. But some species have flipped this logic on its head. We have done so to the greatest extent, where a lack of compassion is now considered a psychological disorder. This "scratch my back and I'll scratch yours" strategy has served us well and is the key to future prosperity. Helping people regardless of their genetic equivalency to us is the ultimate rebellion against selfish genes. By defying the primitive urges of me, me, me, we can flip off the selfish genes and live a life of human nurture, rather than human nature.

I think we're all up to the challenge.

ACKNOWLEDGMENTS

As you should have learned by now, I wrote this book because I had no choice. As you should also have learned by now, nothing happens in a vacuum, so there are many people in this symphony whose horn should be tooted.

First and foremost, I have to thank my parents. Sure, they supplied the genes, but they also provided a healthy environment to satisfy my restless and inquisitive brain. From countless books and records, to my Dataman calculator and Commodore 64, they sacrificed a great deal to nurture and encourage their athlemorphobic kid.

I am in debt to many outstanding teachers and research mentors, especially Drs. William Vail, David Roos, Chuck Smith, and Sherry Queener. They introduced me to the excitement of biomedical research and taught me how to think critically. If it weren't for them, I probably would have made millions selling multiplatinum records and playing hundreds of sold-out shows all over the word with my rock band. So . . . thanks?

Strangely, I have to thank *Toxoplasma gondii*. I've been studying this parasite since 1994, and it introduced me to the idea that things beyond our control affect behavior. During my time in the lab, Carl Zimmer interviewed my research mentor about *Toxoplasma* for his landmark book, *Parasite Rex*. (On p. 118, when Zimmer mentions "Roos's graduate students," that was me!) I figured it might be fun to write a popular science book one day. Jared Diamond is the reason why it took me so long: At a book signing, I told him of my aspirations. Staring at me through empathetic eyes, he sagely advised, "Wait until you get tenure."

Finally, had I not been in Philadelphia at this time, I would not have met Lori, who agreed to team up for the greatest experiment yet: children. Through Colin and Sophia, I was able to study firsthand how genetics plays out.

Indianapolis has a thriving community of scientists and science enthusiasts! I am so grateful for their invitations to leave the lab once in a while to ramble on about fascinating tales in biology. Melanie Fox, who founded Central Indiana Science Outreach (CINSO), invited me to speak at Pint of Science in 2016. I gave a talk called "It's the End of Free Will as We Know It (And I Feel Fine)" that became the basis of this book. Others who provide support, encouragement, and forums to communicate science to the masses include Reba Boyd Wooden (executive director, Center for Inquiry), Rebecca Smith and the Indiana State Museum, Cari Lewis-Tsinovoi (founder of "Books, Booze, and Brains" science book club), and Rufus Cochran (founder of Indiana Science Communication and Education Foundation; March for Science).

I've also been privileged to work with some fantastic people dedicated to the art of science communication, particularly my fellow PLOS SciComm editors, Jason Organ and Krista Hoffmann-Longtin. I owe special thanks to Jason, who endured early drafts of the manuscript and offered many helpful suggestions. I also thank Mark Lasbury, who helped me refine the early ideas I had for this book and co-founded a blogging venture with me called the 'SCOPE, which helped get our writing muscles into shape.

When writing formulaic scientific papers, you rarely encounter writer's block. But in writing a popular science book, I suffered from it on numerous occasions. I have to thank the geniuses at Deviate Brewing for crafting tasty potions that dissolved these cognitive barriers, or at least helped me forget about them for a while.

I count myself very lucky to have Laurie Abkemeier and the team at DeFiore and Company representing me. Laurie has been a tireless champion of this project from day one and patiently helped this rookie draft a book proposal that did not suck. As the book took shape, Laurie helped bring the concepts into focus and weeded out the bad jokes. My talented editors, Hilary Black and Allyson Johnson, operated on the text with CRISPR/Cas9 efficiency and fueled my efforts with their contagious enthusiasm for the subject. I also thank the rest of the team at National Geographic: Melissa Farris (creative director), Nicole Miller (designer), Judith Klein (senior production editor), and Jennifer Thornton (managing editor), as well as Heather McElwain (copy editor).

Last but not least, there would be nothing to write about if not for the curious and hardworking scientists dedicated to advancing human knowledge. It is an honor to discover and assemble the pieces of this magnificent puzzle with you.

SELECT SOURCES

Chapter 1

Borghol, N., M. Suderman, W. McArdle, A. Racine, M. Hallett, M. Pembrey, C. Hertzman, C. Power, and M. Szyf. "Associations With Early-Life Socio-Economic Position in Adult DNA Methylation." *International Journal of Epidemiology* 41, no. 1 (Feb. 2012): 62–74.

Human Microbiome Project, Consortium. "Structure, Function and Diversity of the Healthy Human Microbiome." *Nature* 486, no. 7402 (June 13, 2012): 207–14.

Kioumourtzoglou, M. A., B. A. Coull, E. J. O'Reilly, A. Ascherio, and M. G. Weisskopf. "Association of Exposure to Diethylstilbestrol During Pregnancy With Multigenerational Neurodevelopmental Deficits." *JAMA Pediatrics* 172, no. 7 (July 1, 2018): 670–77.

Lax, S., D. P. Smith, J. Hampton-Marcell, S. M. Owens, K. M. Handley, N. M. Scott, S. M. Gibbons, et al. "Longitudinal Analysis of Microbial Interaction Between Humans and the Indoor Environment." *Science* 345, no. 6200 (Aug. 29, 2014): 1048–52.

Meadow, J. F., A. E. Altrichter, A. C. Bateman, J. Stenson, G. Z. Brown, J. L. Green, and B. J. Bohannan. "Humans Differ in Their Personal Microbial Cloud." *PeerJ* 3 (2015): e1258.

Sender, R., S. Fuchs, and R. Milo. "Revised Estimates for the Number of Human and Bacteria Cells in the Body." *PLoS Biology* 14, no. 8 (Aug. 2016): e1002533.

Simola, D. F., R. J. Graham, C. M. Brady, B. L. Enzmann, C. Desplan, A. Ray, L. J. Zwiebel, et al. "Epigenetic (Re)Programming of Caste-Specific Behavior in the Ant *Camponotus floridanus*." *Science* 351, no. 6268 (Jan. 1, 2016): aac6633.

Chapter 2

Allen, A. L., J. E. McGeary, and J. E. Hayes. "Polymorphisms in TRPV1 and TAS2RS Associate With Sensations From Sampled Ethanol." *Alcoholism: Clinical and Experimental Research* 38, no. 10 (Oct. 2014): 2550–60.

Anderson, E. C., and L. F. Barrett. "Affective Beliefs Influence the Experience of Eating Meat." *PLoS One* 11, no. 8 (2016): e0160424.

Bady, I., N. Marty, M. Dallaporta, M. Emery, J. Gyger, D. Tarussio, M. Foretz, and B. Thorens. "Evidence From Glut2-Null Mice That Glucose Is a Critical Physiological Regulator of Feeding." *Diabetes* 55, no. 4 (Apr. 2006): 988–95.

Basson, M. D., L. M. Bartoshuk, S. Z. Dichello, L. Panzini, J. M. Weiffenbach, and V. B. Duffy. "Association Between 6-N-Propylthiouracil (Prop) Bitterness and Colonic Neoplasms." *Digestive Diseases and Sciences* 50, no. 3 (Mar. 2005): 483–89.

Bayol, S. A., S. J. Farrington, and N. C. Stickland. "A Maternal 'Junk Food' Diet in Pregnancy and Lactation Promotes an Exacerbated Taste for 'Junk Food' and a Greater Propensity for Obesity in Rat Offspring." *British Journal of Nutrition* 98, no. 4 (Oct. 2007): 843–51.

Ceja-Navarro, J. A., F. E. Vega, U. Karaoz, Z. Hao, S. Jenkins, H. C. Lim, P. Kosina, et al. "Gut Microbiota Mediate Caffeine Detoxification in the Primary Insect Pest of Coffee." *Nature Communications* 6 (July 14, 2015): 7618.

Cornelis, M. C., A. El-Sohemy, E. K. Kabagambe, and H. Campos. "Coffee, Cyp1a2 Genotype, and Risk of Myocardial Infarction." *JAMA* 295, no. 10 (Mar. 8, 2006): 1135–41.

Eny, K. M., T. M. Wolever, B. Fontaine-Bisson, and A. El-Sohemy. "Genetic Variant in the Glucose Transporter Type 2 Is Associated With Higher Intakes of Sugars in Two Distinct Populations." *Physiological Genomics* 33, no. 3 (May 13, 2008): 355–60.

Eriksson, N., S. Wu, C. B. Do, A. K. Kiefer, J. Y. Tung, J. L. Mountain, D. A. Hinds, and U. Francke. "A Genetic Variant Near Olfactory Receptor Genes Influences Cilantro Preference." *arXiv.org* (2012).

Hodgson, R. T. "An Examination of Judge Reliability at a Major U.S. Wine Competition." *Journal of Wine Economics* 3, no. 2 (2008): 105–13.

Knaapila, A., L. D. Hwang, A. Lysenko, F. F. Duke, B. Fesi, A. Khoshnevisan, R. S. James, et al. "Genetic Analysis of Chemosensory Traits in Human Twins." *Chemical Senses* 37, no. 9 (Nov. 2012): 869–81.

Marco, A., T. Kisliouk, T. Tabachnik, N. Meiri, and A. Weller. "Overweight and CpG Methylation of the Pomc Promoter in Offspring of High-Fat-Diet-Fed Dams Are Not 'Reprogrammed' by Regular Chow Diet in Rats." *FASEB Journal* 28, no. 9 (Sep. 2014): 4148–57.

McClure, S. M., J. Li, D. Tomlin, K. S. Cypert, L. M. Montague, and P. R. Montague. "Neural Correlates of Behavioral Preference for Culturally Familiar Drinks." *Neuron* 44, no. 2 (Oct. 14, 2004): 379–87.

Mennella, J. A., A. Johnson, and G. K. Beauchamp. "Garlic Ingestion by Pregnant Women Alters the Odor of Amniotic Fluid." *Chemical Senses* 20, no. 2 (Apr. 1995): 207–09.

Munoz-Gonzalez, C., C. Cueva, M. Angeles Pozo-Bayon, and M. Victoria Moreno-Arribas. "Ability of Human Oral Microbiota to Produce Wine Odorant Aglycones From Odourless Grape Glycosidic Aroma Precursors." *Food Chemistry* 187 (Nov. 15, 2015): 112–19.

Pirastu, N., M. Kooyman, M. Traglia, A. Robino, S. M. Willems, G. Pistis, N. Amin, et al. "A Genome-Wide Association Study in Isolated Populations Reveals New Genes Associated to Common Food Likings." *Reviews in Endocrine and Metabolic Disorders* 17, no. 2 (June 2016): 209–19.

Pomeroy, R. "The Legendary Study That Embarrassed Wine Experts Across the Globe." *Real Clear Science,* accessed February 22, 2018, www.realclearscience.com/blog/2014/08/the_most_infamous_study_on_wine_tasting.html.

Rozin, P., L. Millman, and C. Nemeroff. "Operation of the Laws of Sympathetic Magic in Disgust and Other Domains." *Journal of Personality and Social Psychology* 50, no. 4 (1986): 703–12.

Tewksbury, J. J., and G. P. Nabhan. "Seed Dispersal. Directed Deterrence by Capsaicin in Chilies." *Nature* 412, no. 6845 (July 26, 2001): 403–04.

Vani, H. *The Food Babe Way: Break Free From the Hidden Toxins in Your Food and Lose Weight, Look Years Younger, and Get Healthy in Just 21 Days!* New York: Little, Brown and Company, 2015.

Vilanova, C., A. Iglesias, and M. Porcar. "The Coffee-Machine Bacteriome: Biodiversity and Colonisation of the Wasted Coffee Tray Leach." *Scientific Reports* 5 (Nov. 23, 2015): 17163.

Womack, C. J., M. J. Saunders, M. K. Bechtel, D. J. Bolton, M. Martin, N. D. Luden, W. Dunham, and M. Hancock. "The Influence of a Cyp1a2 Polymorphism on the Ergogenic Effects of Caffeine." *Journal of the International Society of Sports Nutrition* 9, no. 1 (Mar. 15, 2012): 7.

Chapter 3

Afshin, A., M. H. Forouzanfar, M. B. Reitsma, P. Sur, K. Estep, A. Lee, et al., and Global Burden of Disease 2015 Obesity Collaborators. "Health Effects of Overweight and Obesity in 195 Countries over 25 Years." *New England Journal of Medicine* 377, no. 1 (July 6, 2017): 13–27.

Ahmed, S. H., K. Guillem, and Y. Vandaele. "Sugar Addiction: Pushing the Drug-Sugar Analogy to the Limit." *Current Opinions in Clinical Nutritition and Metabolic Care* 16, no. 4 (July 2013): 434–39.

Backhed, F., H. Ding, T. Wang, L. V. Hooper, G. Y. Koh, A. Nagy, C. F. Semenkovich, and J. I. Gordon. "The Gut Microbiota as an Environmental Factor That Regulates Fat Storage." *Proceedings of the National Academy of Sciemnces of the United States of America* 101, no. 44 (Nov. 2, 2004): 15718–23.

Barton, W., N. C. Penney, O. Cronin, I. Garcia-Perez, M. G. Molloy, E. Holmes, F. Shanahan, P. D. Cotter, and O. O'Sullivan. "The Microbiome of Professional Athletes Differs From That of More Sedentary Subjects in Composition and Particularly at the Functional Metabolic Level." *Gut* (Mar. 30, 2017).

Blaisdell, A. P., Y. L. Lau, E. Telminova, H. C. Lim, B. Fan, C. D. Fast, D. Garlick, and D. C. Pendergrass. "Food Quality and Motivation: A Refined Low-Fat Diet Induces Obesity and Impairs Performance on a Progressive Ratio Schedule of Instrumental Lever Pressing in Rats." *Physiology & Behavior* 128 (Apr. 10, 2014): 220–25.

Bressa, C., M. Bailen-Andrino, J. Perez-Santiago, R. Gonzalez-Soltero, M. Perez, M. G. Montalvo-Lominchar, J. L. Mate-Munoz, et al. "Differences in Gut Microbiota Profile Between Women With Active Lifestyle and Sedentary Women." *PLoS One* 12, no. 2 (2017): e0171352.

Clement, K., C. Vaisse, N. Lahlou, S. Cabrol, V. Pelloux, D. Cassuto, M. Gourmelen, et al. "A Mutation in the Human Leptin Receptor Gene Causes Obesity and Pituitary Dysfunction." *Nature* 392, no. 6674 (Mar. 26, 1998): 398–401.

De Filippo, C., D. Cavalieri, M. Di Paola, M. Ramazzotti, J. B. Poullet, S. Massart, S. Collini, G. Pieraccini, and P. Lionetti. "Impact of Diet in Shaping Gut Microbiota Revealed by a Comparative Study in Children From Europe and Rural Africa." *Proceedings of the National Academy of Sciences of the United States of America* 107, no. 33 (Aug. 17, 2010): 14691–96.

den Hoed, M., S. Brage, J. H. Zhao, K. Westgate, A. Nessa, U. Ekelund, T. D. Spector, N. J. Wareham, and R. J. Loos. "Heritability of Objectively Assessed Daily Physical Activity and Sedentary Behavior." *American Journal of Clinical Nutrition* 98, no. 5 (Nov. 2013): 1317–25.

Deriaz, O., A. Tremblay, and C. Bouchard. "Non Linear Weight Gain With Long Term Overfeeding in Man." *Obesity Research* 1, no. 3 (May 1993): 179–85.

Derrien, M., C. Belzer, and W. M. de Vos. "*Akkermansia muciniphila* and Its Role in Regulating Host Functions." *Microbial Pathogenesis* 106 (May 2017): 171–81.

Dobson, A. J., M. Ezcurra, C. E. Flanagan, A. C. Summerfield, M. D. Piper, D. Gems, and N. Alic. "Nutritional Programming of Lifespan by Foxo Inhibition on Sugar-Rich Diets." *Cell Reports* 18, no. 2 (Jan. 10, 2017): 299–306.

Dolinoy, D. C., D. Huang, and R. L. Jirtle. "Maternal Nutrient Supplementation Counteracts Bisphenol A-Induced DNA Hypomethylation in Early Development." *Proceedings of the National Academy of the Sciences USA* 104, no. 32 (Aug. 7, 2007): 13056–61.

Donkin, I., S. Versteyhe, L. R. Ingerslev, K. Qian, M. Mechta, L. Nordkap, B. Mortensen, et al. "Obesity and Bariatric Surgery Drive Epigenetic Variation of Spermatozoa in Humans." *Cell Metabolism* 23, no. 2 (Feb. 9, 2016): 369–78.

Everard, A., V. Lazarevic, M. Derrien, M. Girard, G. G. Muccioli, A. M. Neyrinck, S. Possemiers, et al. "Responses of Gut Microbiota and Glucose and Lipid Metabolism to Prebiotics in Genetic Obese and Diet-Induced Leptin-Resistant Mice." *Diabetes* 60, no. 11 (Nov. 2011): 2775–86.

Farooqi, I. S. "Leptin and the Onset of Puberty: Insights From Rodent and Human Genetics." *Seminars in Reproductive Medicine* 20, no. 2 (May 2002): 139–44.

Grimm, E. R., and N. I. Steinle. "Genetics of Eating Behavior: Established and Emerging Concepts." *Nutrition Reviews* 69, no. 1 (Jan. 2011): 52–60.

Hehemann, J. H., G. Correc, T. Barbeyron, W. Helbert, M. Czjzek, and G. Michel. "Transfer of Carbohydrate-Active Enzymes From Marine Bacteria to Japanese Gut Microbiota." *Nature* 464, no. 7290 (Apr. 8, 2010): 908–12.

Johnson, R. K., L. J. Appel, M. Brands, B. V. Howard, M. Lefevre, R. H. Lustig, F. Sacks, et al. "Dietary Sugars Intake and Cardiovascular Health: A Scientific Statement From the American Heart Association." *Circulation* 120, no. 11 (Sep. 15, 2009): 1011–20.

Jumpertz, R., D. S. Le, P. J. Turnbaugh, C. Trinidad, C. Bogardus, J. I. Gordon, and J. Krakoff. "Energy-Balance Studies Reveal Associations Between Gut Microbes, Caloric Load, and Nutrient Absorption in Humans." *American Journal of Clinical Nutrition* 94, no. 1 (July 2011): 58–65.

Levine, J. A. "Solving Obesity Without Addressing Poverty: Fat Chance." *Journal of Hepatology* 63, no. 6 (Dec. 2015): 1523–24.

Ley, R. E., F. Backhed, P. Turnbaugh, C. A. Lozupone, R. D. Knight, and J. I. Gordon. "Obesity Alters Gut Microbial Ecology." *Proceedings of the National Academy of the Sciences USA* 102, no. 31 (Aug. 2, 2005): 11070–05.

Loos, R. J., C. M. Lindgren, S. Li, E. Wheeler, J. H. Zhao, I. Prokopenko, M. Inouye, et al. "Common Variants Near Mc4r Are Associated With Fat Mass, Weight and Risk of Obesity." *Nature Genetics* 40, no. 6 (June 2008): 768–75.

Mann, Traci. *Secrets From the Eating Lab*. Harper Wave, 2015.

Moss, Michael. *Salt Sugar Fat: How the Food Giants Hooked Us*. New York: Random House, 2013.

Ng, S. F., R. C. Lin, D. R. Laybutt, R. Barres, J. A. Owens, and M. J. Morris. "Chronic High-Fat Diet in Fathers Programs Beta-Cell Dysfunction in Female Rat Offspring." *Nature* 467, no. 7318 (Oct. 21, 2010): 963–66.

O'Rahilly, S. "Life Without Leptin." *Nature* 392, no. 6674 (Mar. 26, 1998): 330–31.

Pelleymounter, M. A., M. J. Cullen, M. B. Baker, R. Hecht, D. Winters, T. Boone, and F. Collins. "Effects of the Obese Gene Product on Body Weight Regulation in Ob/Ob Mice." *Science* 269, no. 5223 (July 28, 1995): 540–43.

Puhl, R., and Y. Suh. "Health Consequences of Weight Stigma: Implications for Obesity Prevention and Treatment." *Current Obesity Reports* 4, no. 2 (June 2015): 182–90.

Ridaura, V. K., J. J. Faith, F. E. Rey, J. Cheng, A. E. Duncan, A. L. Kau, N. W. Griffin, et al. "Gut Microbiota From Twins Discordant for Obesity Modulate Metabolism in Mice." *Science* 341, no. 6150 (Sep. 6, 2013): 1241214.

Roberts, M. D., J. D. Brown, J. M. Company, L. P. Oberle, A. J. Heese, R. G. Toedebusch, K. D. Wells, et al. "Phenotypic and Molecular Differences Between Rats Selectively Bred to Voluntarily Run High Vs. Low Nightly Distances." *American Journal of Physiology-Regulatory Integrative and Comparative Physiology* 304, no. 11 (June 1, 2013): R1024–35.

Schulz, L. O., and L. S. Chaudhari. "High-Risk Populations: The Pimas of Arizona and Mexico." *Current Obesity Reports* 4, no. 1 (Mar. 2015): 92–98.

Shadiack, A. M., S. D. Sharma, D. C. Earle, C. Spana, and T. J. Hallam. "Melanocortins in the Treatment of Male and Female Sexual Dysfunction." *Current Topics in Medicinal Chemistry* 7, no. 11 (2007): 1137–44.

Stice, E., S. Spoor, C. Bohon, and D. M. Small. "Relation Between Obesity and Blunted Striatal Response to Food Is Moderated by Taqia A1 Allele." *Science* 322, no. 5900 (Oct. 17, 2008): 449–52.

Trogdon, J. G., E. A. Finkelstein, C. W. Feagan, and J. W. Cohen. "State- and Payer-Specific Estimates of Annual Medical Expenditures Attributable to Obesity." *Obesity* (Silver Spring) 20, no. 1 (Jan. 2012): 214–20.

Trompette, A., E. S. Gollwitzer, K. Yadava, A. K. Sichelstiel, N. Sprenger, C. Ngom-Bru, C. Blanchard, et al. "Gut Microbiota Metabolism of Dietary Fiber Influences Allergic Airway Disease and Hematopoiesis." *Nature Medicine* 20, no. 2 (Feb. 2014): 159–66.

Turnbaugh, P. J., M. Hamady, T. Yatsunenko, B. L. Cantarel, A. Duncan, R. E. Ley, M. L. Sogin, et al. "A Core Gut Microbiome in Obese and Lean Twins." *Nature* 457, no. 7228 (Jan. 22, 2009): 480–84.

Turnbaugh, P. J., R. E. Ley, M. A. Mahowald, V. Magrini, E. R. Mardis, and J. I. Gordon. "An Obesity-Associated Gut Microbiome With Increased Capacity for Energy Harvest." *Nature* 444, no. 7122 (Dec. 21, 2006): 1027–31.

Voisey, J., and A. van Daal. "Agouti: From Mouse to Man, From Skin to Fat." *Pigment Cell & Melanoma Research* 15, no. 1 (Feb. 2002): 10–18.

Wang L., S. Gillis-Smith, Y. Peng, J. Zhang, X. Chen, C. D. Salzman, N. J. Ryba, and C. S. Zuker. "The Coding of Valence and Identity in the Mammalian Taste System." *Nature* 558, no. 7708 (June 2018): 127–31.

Yang, N., D. G. MacArthur, J. P. Gulbin, A. G. Hahn, A. H. Beggs, S. Easteal, and K. North. "Actn3 Genotype Is Associated With Human Elite Athletic Performance." *American Journal of Human Genetics* 73, no. 3 (Sep. 2003): 627–31.

Zhang, X., and A. N. van den Pol. "Rapid Binge-Like Eating and Body Weight Gain Driven by Zona Incerta GABA Neuron Activation." *Science* 356, no. 6340 (May 26, 2017): 853–59.

Chapter 4

Anstee, Q. M., S. Knapp, E. P. Maguire, A. M. Hosie, P. Thomas, M. Mortensen, R. Bhome, et al. "Mutations in the Gabrb1 Gene Promote Alcohol Consumption Through Increased Tonic Inhibition." *Nature Communications* 4 (2013): 2816.

Bercik, P., E. Denou, J. Collins, W. Jackson, J. Lu, J. Jury, Y. Deng, et al. "The Intestinal Microbiota Affect Central Levels of Brain-Derived Neurotropic Factor and Behavior in Mice." *Gastroenterology* 141, no. 2 (Aug. 2011): 599–609.e3.

Dick, D. M., H. J. Edenberg, X. Xuei, A. Goate, S. Kuperman, M. Schuckit, R. Crowe, et al. "Association of Gabrg3 With Alcohol Dependence." *Alcoholism: Clinical and Experimental Research* 28, no. 1 (Jan. 2004): 4–9.

DiNieri, J. A., X. Wang, H. Szutorisz, S. M. Spano, J. Kaur, P. Casaccia, D. Dow-Edwards, and Y. L. Hurd. "Maternal Cannabis Use Alters Ventral Striatal Dopamine D2 Gene Regulation in the Offspring." *Biological Psychiatry* 70, no. 8 (Oct. 15, 2011): 763–69.

Egervari, G., J. Landry, J. Callens, J. F. Fullard, P. Roussos, E. Keller, and Y. L. Hurd. "Striatal H3k27 Acetylation Linked to Glutamatergic Gene Dysregulation in Human Heroin Abusers Holds Promise as Therapeutic Target." *Biological Psychiatry* 81, no. 7 (Apr. 1, 2017): 585–94.

Finkelstein, E. A., K. W. Tham, B. A. Haaland, and A. Sahasranaman. "Applying Economic Incentives to Increase Effectiveness of an Outpatient Weight Loss Program (Trio): A Randomized Controlled Trial." *Social Science & Medicine* 185 (July 2017): 63–70.

Flagel, S. B., S. Chaudhury, M. Waselus, R. Kelly, S. Sewani, S. M. Clinton, R. C. Thompson, S. J. Watson, Jr., and H. Akil. "Genetic Background and Epigenetic Modifications in the Core of the Nucleus Accumbens Predict Addiction-Like

Behavior in a Rat Model." *Proceedings of the National Academy of the Sciences USA* 113, no. 20 (May 17. 2016): E2861–70.

Flegr, J., and R. Kuba. "The Relation of *Toxoplasma* Infection and Sexual Attraction to Fear, Danger, Pain, and Submissiveness." *Evolutionary Psychology* 14, no. 3 (2016).

Flegr, J., M. Preiss, J. Klose, J. Havlicek, M. Vitakova, and P. Kodym. "Decreased Level of Psychobiological Factor Novelty Seeking and Lower Intelligence in Men Latently Infected With the Protozoan Parasite *Toxoplasma gondii* Dopamine, a Missing Link Between Schizophrenia and Toxoplasmosis?" *Biological Psychology* 63, no. 3 (July 2003): 253–68.

""Frandsen, M. "Why We Should Pay People to Stop Smoking." theconversation.com/why-we-should-pay-people-to-stop-smoking-84058.

Giordano, G. N., H. Ohlsson, K. S. Kendler, K. Sundquist, and J. Sundquist. "Unexpected Adverse Childhood Experiences and Subsequent Drug Use Disorder: A Swedish Population Study (1995–2011)." *Addiction* 109, no. 7 (July 2014): 1119–27.

Kippin, T. E., J. C. Campbell, K. Ploense, C. P. Knight, and J. Bagley. "Prenatal Stress and Adult Drug-Seeking Behavior: Interactions With Genes and Relation to Nondrug-Related Behavior." *Advances in Neurobiology* 10 (2015): 75–100.

Koepp, M. J., R. N. Gunn, A. D. Lawrence, V. J. Cunningham, A. Dagher, T. Jones, D. J. Brooks, C. J. Bench, and P. M. Grasby. "Evidence for Striatal Dopamine Release During a Video Game." *Nature* 393, no. 6682 (May 21 1998): 266–68.

Kreek, M. J., D. A. Nielsen, E. R. Butelman, and K. S. LaForge. "Genetic Influences on Impulsivity, Risk Taking, Stress Responsivity and Vulnerability to Drug Abuse and Addiction." *Nature Neuroscience* 8, no. 11 (Nov. 2005): 1450–57.

Leclercq, S., S. Matamoros, P. D. Cani, A. M. Neyrinck, F. Jamar, P. Starkel, K. Windey, et al. "Intestinal Permeability, Gut-Bacterial Dysbiosis, and Behavioral Markers of Alcohol-Dependence Severity." *Proceedings of the National Academy of the Sciences USA* 111, no. 42 (Oct. 21, 2014): E4485–93.

Matthews, L. J., and P. M. Butler. "Novelty-Seeking DRD4 Polymorphisms Are Associated With Human Migration Distance Out-of-Africa After Controlling for Neutral Population Gene Structure." *American Journal of Physical Anthropology* 145, no. 3 (July 2011): 382–89.

Mohammad, Akikur. *The Anatomy of Addiction: What Science and Research Tell Us About the True Causes, Best Preventive Techniques, and Most Successful Treatments*. New York: TarcherPerigee, 2016.

Osbourne, Ozzy. *Trust Me, I'm Dr. Ozzy: Advice from Rock's Ultimate Survivor*. New York: Grand Central Publishing, 2011.

Peng, Y., H. Shi, X. B. Qi, C. J. Xiao, H. Zhong, R. L. Ma, and B. Su. "The ADH1B Arg47His Polymorphism in East Asian Populations and Expansion of Rice Domestication in History." *BMC Evolutionary Biology* 10 (Jan. 20, 2010): 15.

Peters, S., and E. A. Crone. "Increased Striatal Activity in Adolescence Benefits Learning." *Nature Communications* 8, no. 1 (Dec. 19, 2017): 1983.

Ptacek, R., H. Kuzelova, and G. B. Stefano. "Dopamine D4 Receptor Gene DRD4 and Its Association With Psychiatric Disorders." *Medical Science Monitor* 17, no. 9 (Sep. 2011): RA215–20.

Repunte-Canonigo, V., M. Herman, T. Kawamura, H. R. Kranzler, R. Sherva, J. Gelernter, L. A. Farrer, M. Roberto, and P. P. Sanna. "Nf1 Regulates Alcohol Dependence-Associated Excessive Drinking and Gamma-Aminobutyric Acid Release in the Central Amygdala in Mice and Is Associated With Alcohol Dependence in Humans." *Biological Psychiatry* 77, no. 10 (May 15, 2015): 870–79.

Reynolds, Gretchen. "The Genetics of Being a Daredevil." *New York Times,* well .blogs.nytimes.com/2014/02/19/the-genetics-of-being-a-daredevil/?_r=0.

Schumann, G., C. Liu, P. O'Reilly, H. Gao, P. Song, B. Xu, B. Ruggeri, et al. "KLB Is Associated With Alcohol Drinking, and Its Gene Product Beta-Klotho Is Necessary for FGF21 Regulation of Alcohol Preference." *Proceedings of the National Academy of the Sciences USA* 113, no. 50 (Dec. 13, 2016): 14372–77.

Stoel, R. D., E. J. De Geus, and D. I. Boomsma. "Genetic Analysis of Sensation Seeking With an Extended Twin Design." *Behavior Genetics* 36, no. 2 (Mar. 2006): 229–37.

Substance Abuse and Mental Health Services Administration. "Substance Use and Dependence Following Initiation of Alcohol of Illicit Drug Use"," *The NSDUH Report,* Rockville, MD, 2008.

Sutterland, A. L., G. Fond, A. Kuin, M. W. Koeter, R. Lutter, T. van Gool, R. Yolken, et al. "Beyond the Association. *Toxoplasma gondii* in Schizophrenia, Bipolar Disorder, and Addiction: Systematic Review and Meta-Analysis." *Acta Psychiatrica Scandinavica* 132, no. 3 (Sep. 2015): 161–79.

Szalavitz, Maia. *Unbroken Brain: A Revolutionary New Way of Understanding Addiction.* New York: St. Martin's Press, 2016.

Tikkanen, R., J. Tiihonen, M. R. Rautiainen, T. Paunio, L. Bevilacqua, R. Panarsky, D. Goldman, and M. Virkkunen. "Impulsive Alcohol-Related Risk-Behavior and Emotional Dysregulation Among Individuals With a Serotonin 2b Receptor Stop Codon." *Translational Psychiatry* 5 (Nov. 17, 2015): e681.

Vallee, M., S. Vitiello, L. Bellocchio, E. Hebert-Chatelain, S. Monlezun, E. Martin-Garcia, F. Kasanetz, et al. "Pregnenolone Can Protect the Brain From Cannabis Intoxication." *Science* 343, no. 6166 (Jan. 3, 2014): 94–98.

Webb, A., P. A. Lind, J. Kalmijn, H. S. Feiler, T. L. Smith, M. A. Schuckit, and K. Wilhelmsen. "The Investigation Into CYP2E1 in Relation to the Level of Response to Alcohol Through a Combination of Linkage and Association Analysis." *Alcoholism: Clinical and Experimental Research* 35, no. 1 (Jan. 2011): 10–18.

Chapter 5

Aldwin, C. M., Y. J. Jeong, H. Igarashi, and A. Spiro. "Do Hassles and Uplifts Change With Age? Longitudinal Findings From the Va Normative Aging Study." *Psychology and Aging* 29, no. 1 (Mar. 2014): 57–71.

Amin, N., N. M. Belonogova, O. Jovanova, R. W. Brouwer, J. G. van Rooij, M. C. van den Hout, G. R. Svishcheva, et al. "Nonsynonymous Variation in NKPD1 Increases Depressive Symptoms in European Populations." *Biological Psychiatry* 81, no. 8 (Apr. 15, 2017): 702–07.

Bravo, J. A., P. Forsythe, M. V. Chew, E. Escaravage, H. M. Savignac, T. G. Dinan, J. Bienenstock, and J. F. Cryan. "Ingestion of *Lactobacillus* Strain Regulates Emotional Behavior and Central Gaba Receptor Expression in a Mouse Via the Vagus Nerve." *Proceedings of the National Academy of the Sciences USA* 108, no. 38 (Sep. 20, 2011): 16050–55.

Brickman, P., D. Coates, and R. Janoff-Bulman. "Lottery Winners and Accident Victims: Is Happiness Relative?" *Journal of Personality and Social Psychology* 36, no. 8 (Aug. 1978): 917–27.

Cameron, N. M., D. Shahrokh, A. Del Corpo, S. K. Dhir, M. Szyf, F. A. Champagne, and M. J. Meaney. "Epigenetic Programming of Phenotypic Variations in Reproductive Strategies in the Rat Through Maternal Care." *Journal of Neuroendocrinology* 20, no. 6 (June 2008): 795–801.

Caspi, A., K. Sugden, T. E. Moffitt, A. Taylor, I. W. Craig, H. Harrington, J. McClay, et al. "Influence of Life Stress on Depression: Moderation by a Polymorphism in the 5-HTT Gene." *Science* 301, no. 5631 (July 18, 2003): 386–89.

Chiao, J. Y., and K. D. Blizinsky. "Culture-Gene Coevolution of Individualism-Collectivism and the Serotonin Transporter Gene." *Proceedings of the Royal Society of London B: Biological Sciences* 277, no. 1681 (Feb. 22, 2010): 529–37.

Claesson, M. J., S. Cusack, O. O'Sullivan, R. Greene-Diniz, H. de Weerd, E. Flannery, J. R. Marchesi, et al. "Composition, Variability, and Temporal Stability of the Intestinal Microbiota of the Elderly." *Proceedings of the National Academy of the Sciences USA* 108 Suppl. 1 (Mar. 15, 2011): 4586–91.

Claesson, M. J., I. B. Jeffery, S. Conde, S. E. Power, E. M. O'Connor, S. Cusack, H. M. Harris, et al. "Gut Microbiota Composition Correlates With Diet and Health in the Elderly." *Nature* 488, no. 7410 (Aug. 9, 2012): 178–84.

Converge Consortium. "Sparse Whole-Genome Sequencing Identifies Two Loci for Major Depressive Disorder." *Nature* 523, no. 7562 (July 30, 2015): 588–91.

Cordell, B., and J. McCarthy. "A Case Study of Gut Fermentation Syndrome (Auto-Brewery) With Saccharomyces Cerevisiae as the Causative Organism." " *International Journal of Clinical Medicine* 4 (2013): 309–12.

Dreher J. C., S. Dunne S, A. Pazderska, T. Frodl, J. J. Nolan, and J. P. O'Doherty. "Testosterone Causes Both Prosocial and Antisocial Status-Enhancing Behaviors in Human Males." *Proceedings of the National Academy of the Sciences USA* 113, no. 41 (Oct. 11, 2016): 11633–38.

Ford, B. Q., M. Tamir, T. T. Brunye, W. R. Shirer, C. R. Mahoney, and H. A. Taylor. "Keeping Your Eyes on the Prize: Anger and Visual Attention to Threats and Rewards." *Psychological Science* 21, no. 8 (Aug. 2010): 1098–105.

Gruber, J., I. B. Mauss, and M. Tamir. "A Dark Side of Happiness? How, When, and Why Happiness Is Not Always Good." *Perspectives on Psychological Science* 6, no. 3 (May 2011): 222–33.

Guccione, Bob. "Fanfare for the Common Man: Who Is John Mellencamp?" *SPIN,* 1992.

Hing, B., C. Gardner, and J. B. Potash. "Effects of Negative Stressors on DNA Methylation in the Brain: Implications for Mood and Anxiety Disorders." *American Journal of Medical Genetics B: Neuropsychiatric Genetics* 165B, no. 7 (Oct. 2014): 541–54.

Hyde, C. L., M. W. Nagle, C. Tian, X. Chen, S. A. Paciga, J. R. Wendland, J. Y. Tung, et al. "Identification of 15 Genetic Loci Associated With Risk of Major Depression in Individuals of European Descent." *Nature Genetics* 48, no. 9 (Sep. 2016): 1031–36.

Jansson-Nettelbladt, E., S. Meurling, B. Petrini, and J. Sjolin. "Endogenous Ethanol Fermentation in a Child With Short Bowel Syndrome." *Acta Paediatrica* 95, no. 4 (Apr. 2006): 502–04.

Kaufman, J., B. Z. Yang, H. Douglas-Palumberi, S. Houshyar, D. Lipschitz, J. H. Krystal, and J. Gelernter. "Social Supports and Serotonin Transporter Gene Moderate Depression in Maltreated Children." *Proceedings of the National Academy of the Sciences USA* 101, no. 49 (Dec. 7, 2004): 17316–21.

Kelly, J. R., Y. Borre, O' Brien C, E. Patterson, S. El Aidy, J. Deane, P. J. Kennedy, et al. "Transferring the Blues: Depression-Associated Gut Microbiota Induces Neurobehavioural Changes in the Rat." *Journal of Psychiatric Research* 82 (Nov. 2016): 109–18.

Kim A., and S. J. Maglio. "Vanishing Time in the Pursuit of Happiness." *Psychonomic Bulletin and Review* 25, no. 4 (Aug. 2018): 1337–42.

LaMotte, S. "Woman Claims Her Body Brews Alcohol, Has DUI Charge Dismissed." *CNN,* www.cnn.com/2015/12/31/health/auto-brewery-syndrome-dui-womans-body-brews-own-alcohol/index.html.

Lohoff, F. W. "Overview of the Genetics of Major Depressive Disorder." *Current Psychiatry Reports* 12, no. 6 (Dec. 2010): 539–46.

McGowan, P. O., A. Sasaki, A. C. D'Alessio, S. Dymov, B. Labonte, M. Szyf, G. Turecki, and M. J. Meaney. "Epigenetic Regulation of the Glucocorticoid Receptor in Human Brain Associates With Childhood Abuse." *Nature Neuroscience* 12, no. 3 (Mar. 2009): 342–48.

Messaoudi, M., R. Lalonde, N. Violle, H. Javelot, D. Desor, A. Nejdi, J. F. Bisson, et al. "Assessment of Psychotropic-Like Properties of a Probiotic Formulation *(Lactobacillus helveticus* R0052 and *Bifidobacterium longum* R0175) in Rats and Human Subjects." *British Journal of Nutrition* 105, no. 5 (Mar. 2011): 755–64.

Minkov, M., and M. H. Bond. "A Genetic Component to National Differences in Happiness." *Journal of Happiness Studies* 18, no. 2 (2017): 321–40.

Moll, J., F. Krueger, R. Zahn, M. Pardini, R. de Oliveira-Souza, and J. Grafman. "Human Fronto-Mesolimbic Networks Guide Decisions About Charitable Donation." *Proceedings of the National Academy of the Sciences USA* 103, no. 42 (Oct. 17, 2006): 15623–28.

Naumova, O. Y., M. Lee, R. Koposov, M. Szyf, M. Dozier, and E. L. Grigorenko. "Differential Patterns of Whole-Genome DNA Methylation in Institutionalized Children and Children Raised by Their Biological Parents." *Development and Psychopathology* 24, no. 1 (Feb. 2012): 143–55.

Nesse, R. M. "Natural Selection and the Elusiveness of Happiness." *Philosophical Transactions of the Royal Society London B: Biological Sciences* 359, no. 1449 (Sep. 29, 2004): 1333–47.

Okbay, A., B. M. Baselmans, J. E. De Neve, P. Turley, M. G. Nivard, M. A. Fontana, S. F. Meddens, et al. "Genetic Variants Associated With Subjective Well-Being, Depressive Symptoms, and Neuroticism Identified Through Genome-Wide Analyses." *Nature Genetcis* 48, no. 6 (June 2016): 624–33.

Pena, C. J., H. G. Kronman, D. M. Walker, H. M. Cates, R. C. Bagot, I. Purushothaman, O. Issler, et al. "Early Life Stress Confers Lifelong Stress Susceptibility in Mice Via Ventral Tegmental Area Otx2." *Science* 356, no. 6343 (June 16, 2017): 1185–88.

Pronto, E., and Pswald A. J. "National Happiness and Genetic Distance: A Cautious Exploration." ftp.iza.org/dp8300.pdf.

Romens, S. E., J. McDonald, J. Svaren, and S. D. Pollak. "Associations Between Early Life Stress and Gene Methylation in Children." *Child Development* 86, no. 1 (Jan.–Feb. 2015): 303–09.

Rosenbaum, J. T. "The E. Coli Made Me Do It." *The New Yorker,* www.newyorker.com/tech/elements/the-e-coli-made-me-do-it.

Singer, Peter. *The Expanding Circle: Ethics and Sociobiology*. Princeton, NJ: Princeton University Press, 1981.

Steenbergen, L., R. Sellaro, S. van Hemert, J. A. Bosch, and L. S. Colzato. "A Randomized Controlled Trial to Test the Effect of Multispecies Probiotics on Cognitive Reactivity to Sad Mood." *Brain, Behavior, and Immunity* 48 (Aug. 2015): 258–64.

Sudo, N., Y. Chida, Y. Aiba, J. Sonoda, N. Oyama, X. N. Yu, C. Kubo, and Y. Koga. "Postnatal Microbial Colonization Programs the Hypothalamic-Pituitary-Adrenal System for Stress Response in Mice." *Journal of Physiology* 558, no. Pt. 1 (July 1, 2004): 263–75.

Sullivan, P. F., M. C. Neale, and K. S. Kendler. "Genetic Epidemiology of Major Depression: Review and Meta-Analysis." *American Journal of Psychiatry* 157, no. 10 (Oct. 2000): 1552–62.

Swartz, J. R., A. R. Hariri, and D. E. Williamson. "An Epigenetic Mechanism Links Socioeconomic Status to Changes in Depression-Related Brain Function in High-Risk Adolescents." *Molecular Psychiatry* 22, no. 2 (Feb. 2017): 209–14.

Tillisch, K., J. Labus, L. Kilpatrick, Z. Jiang, J. Stains, B. Ebrat, D. Guyonnet, et al. "Consumption of Fermented Milk Product With Probiotic Modulates Brain Activity." *Gastroenterology* 144, no. 7 (June 2013): 1394–401.e4.

World Health Organization. "Depression." www.who.int/mediacentre/factsheets/fs369/en.

Zhang, L., A. Hirano, P. K. Hsu, C. R. Jones, N. Sakai, M. Okuro, T. McMahon, et al. "A PERIOD3 Variant Causes a Circadian Phenotype and Is Associated With a Seasonal Mood Trait." *Proceedings of the National Academy of the Sciences USA* 113, no. 11 (Mar. 15, 2016): E1536–44.

Chapter 6

Aizer, A., and J. Currie. "Lead and Juvenile Delinquency: New Evidence From Linked Birth, School and Juvenile Detention Records." National Bureau of Economic Research, www.nber.org/papers/w23392.

Arrizabalaga, G., and B. Sullivan. "Common Parasite Could Manipulate Our Behavior." Scientific American MIND, www.scientificamerican.com/article/common-parasite-could-manipulate-our-behavior.

Berdoy, M., J. P. Webster, and D. W. Macdonald. "Fatal Attraction in Rats Infected With *Toxoplasma gondii.*" *Proceedings of the Royal Society: Biological Sciences* 267, no. 1452 (Aug. 7, 2000): 1591–94.

Bjorkqvist, K. "Gender Differences in Aggression." *Current Opinion in Psychology* 19 (Feb. 2018): 39–42.

Brunner, H. G., M. Nelen, X. O. Breakefield, H. H. Ropers, and B. A. van Oost. "Abnormal Behavior Associated With a Point Mutation in the Structural Gene for Monoamine Oxidase A." *Science* 262, no. 5133 (Oct. 22, 1993): 578–80.

Burgess, E. E., M. D. Sylvester, K. E. Morse, F. R. Amthor, S. Mrug, K. L. Lokken, M. K. Osborn, T. Soleymani, and M. M. Boggiano. "Effects of Transcranial Direct Current Stimulation (Tdcs) on Binge Eating Disorder." *International Journal of Eating Disorders* 49, no. 10 (Oct. 2016): 930–36.

Burt, S. A. "Are There Meaningful Etiological Differences Within Antisocial Behavior? Results of a Meta-Analysis." *Clinical Psychology Review* 29, no. 2 (Mar. 2009): 163–78.

Cahalan, Susannah. *Brain on Fire: My Month of Madness.* New York: Simon & Schuster, 2013.

Cases, O., I. Seif, J. Grimsby, P. Gaspar, K. Chen, S. Pournin, U. Muller, et al. "Aggressive Behavior and Altered Amounts of Brain Serotonin and Norepinephrine in Mice Lacking Maoa." *Science* 268, no. 5218 (June 23, 1995): 1763–66.

Caspi, A., J. McClay, T. E. Moffitt, J. Mill, J. Martin, I. W. Craig, A. Taylor, and R. Poulton. "Role of Genotype in the Cycle of Violence in Maltreated Children." *Science* 297, no. 5582 (Aug. 2, 2002): 851–54.

Chen, H., D. S. Pine, M. Ernst, E. Gorodetsky, S. Kasen, K. Gordon, D. Goldman, and P. Cohen. "The Maoa Gene Predicts Happiness in Women." *Progress in Neuropsychopharmacology & Biological Psychiatry* 40 (Jan. 10, 2013): 122–25.

Coccaro, E. F., R. Lee, M. W. Groer, A. Can, M. Coussons-Read, and T. T. Postolache. "*Toxoplasma gondii* Infection: Relationship With Aggression in Psychiatric Subjects." *Journal of Clinical Psychiatry* 77, no. 3 (Mar. 2016): 334–41.

Crockett, M. J., L. Clark, G. Tabibnia, M. D. Lieberman, and T. W. Robbins. "Serotonin Modulates Behavioral Reactions to Unfairness." *Science* 320, no. 5884 (June 27, 2008): 1739.

Dalmau, J., E. Tuzun, H. Y. Wu, J. Masjuan, J. E. Rossi, A. Voloschin, J. M. Baehring, et al. "Paraneoplastic Anti-N-Methyl-D-Aspartate Receptor Encephalitis Associated With Ovarian Teratoma." *Annals of Neurology* 61, no. 1 (Jan. 2007): 25–36.

Dias, B. G., and K. J. Ressler. "Parental Olfactory Experience Influences Behavior and Neural Structure in Subsequent Generations." *Nature Neuroscience* 17, no. 1 (Jan. 2014): 89–96.

Faiola, A. "A Modern Pope Gets Old School on the Devil." *The Washington Post,* www.washingtonpost.com/world/a-modern-pope-gets-old-school-on-the-devil/2014/05/10/f56a9354-1b93-4662-abbb-d877e49f15ea_story.html?utm_term=.8a6c61629cd5.

Feigenbauma, J.J., and C. Muller. "Lead Exposure and Violent Crime in the Early Twentieth Century." *Explorations in Economic History* 62 (2016): 51–86.

Flegr, J., J. Havlicek, P. Kodym, M. Maly, and Z. Smahel. "Increased Risk of Traffic Accidents in Subjects With Latent Toxoplasmosis: A Retrospective Case-Control Study." *BMC Infectious Diseases* 2 (July 2, 2002): 11.

Gatzke-Kopp, L. M., and T. P. Beauchaine. "Direct and Passive Prenatal Nicotine Exposure and the Development of Externalizing Psychopathology." *Child Psychiatry and Human Development* 38, no. 4 (Dec. 2007): 255–69.

Gogos, J. A., M. Morgan, V. Luine, M. Santha, S. Ogawa, D. Pfaff, and M. Karayiorgou. "Catechol-O-Methyltransferase-Deficient Mice Exhibit Sexually Dimorphic Changes in Catecholamine Levels and Behavior." *Proceedings of the National Academy of the Sciences USA* 95, no. 17 (Aug. 18, 1998): 9991–96.

Gunduz-Cinar, O., M. N. Hill, B. S. McEwen, and A. Holmes. "Amygdala FAAH and Anandamide: Mediating Protection and Recovery from Stress." *Trends in Pharmacological Sciences* 34, no. 11 (Nov. 2013): 637–44.

Hawthorne, M. "Studies Link Childhood Lead Exposure, Violent Crime." *Chicago Tribune,* www.chicagotribune.com/news/ct-lead-poisoning-science-met-20150605-story.html.

Heijmans, B. T., E. W. Tobi, A. D. Stein, H. Putter, G. J. Blauw, E. S. Susser, P. E. Slagboom, and L. H. Lumey. "Persistent Epigenetic Differences Associated With Prenatal Exposure to Famine in Humans." *Proceedings of the National Academy of the Sciences USA* 105, no. 44 (Nov. 4, 2008): 17046–49.

Hibbeln, J. R., J. M. Davis, C. Steer, P. Emmett, I. Rogers, C. Williams, and J. Golding. "Maternal Seafood Consumption in Pregnancy and Neurodevelopmental Outcomes in Childhood (Alspac Study): An Observational Cohort Study." *Lancet* 369, no. 9561 (Feb. 17, 2007): 578–85.

Hodges, L. M., A. J. Fyer, M. M. Weissman, M. W. Logue, F. Haghighi, O. Evgrafov, A. Rotondo, J. A. Knowles, and S. P. Hamilton. "Evidence for Linkage and Association of GABRB3 and GABRA5 to Panic Disorder." *Neuropsychopharmacology* 39, no. 10 (Sep. 2014): 2423–31.

Hunter, P. "The Psycho Gene." *EMBO Reports* 11, no. 9 (Sep. 2010): 667–69.

Ivorra, C., M. F. Fraga, G. F. Bayon, A. F. Fernandez, C. Garcia-Vicent, F. J. Chaves, J. Redon, and E. Lurbe. "DNA Methylation Patterns in Newborns Exposed to Tobacco in Utero." *Journal of Translational Medicine* 13 (Jan. 27, 2015): 25.

Kelly, S. J., N. Day, and A. P. Streissguth. "Effects of Prenatal Alcohol Exposure on Social Behavior in Humans and Other Species." *Neurotoxicology and Teratology* 22, no. 2 (Mar.–Apr. 2000): 143–49.

Li, Y., C. Xie, S. K. Murphy, D. Skaar, M. Nye, A. C. Vidal, K. M. Cecil, et al. "Lead Exposure During Early Human Development and DNA Methylation of Imprinted Gene Regulatory Elements in Adulthood." *Environmental Health Perspectives* 124, no. 5 (May 2016): 666–73.

Mednick, S. A., W. F. Gabrielli, Jr., and B. Hutchings. "Genetic Influences in Criminal Convictions: Evidence From an Adoption Cohort." *Science* 224, no. 4651 (May 25, 1984): 891–94.

Neugebauer, R., H. W. Hoek, and E. Susser. "Prenatal Exposure to Wartime Famine and Development of Antisocial Personality Disorder in Early Adulthood." *JAMA* 282, no. 5 (Aug. 4, 1999): 455–62.

Ouellet-Morin, I., C. C. Wong, A. Danese, C. M. Pariante, A. S. Papadopoulos, J. Mill, and L. Arseneault. "Increased Serotonin Transporter Gene (Sert) DNA Methylation Is Associated With Bullying Victimization and Blunted Cortisol Response to Stress in Childhood: A Longitudinal Study of Discordant Monozygotic Twins." *Psychological Medicine* 43, no. 9 (Sep. 2013): 1813–23.

Ouko, L. A., K. Shantikumar, J. Knezovich, P. Haycock, D. J. Schnugh, and M. Ramsay. "Effect of Alcohol Consumption on CpG Methylation in the Differentially Methylated Regions of H19 and IG-DMR in Male Gametes: Implications for Fetal Alcohol Spectrum Disorders." *Alcoholism: Clinical and Experimental Research* 33, no. 9 (Sep. 2009): 1615–27.

Raine, A., J. Portnoy, J. Liu, T. Mahoomed, and J. R. Hibbeln. "Reduction in Behavior Problems With Omega-3 Supplementation in Children Aged 8–16 Years: A Randomized, Double-Blind, Placebo-Controlled, Stratified, Parallel-Group Trial." *Journal of Child Psychology and Psychiatry* 56, no. 5 (May 2015): 509–20.

Ramboz, S., F. Saudou, D. A. Amara, C. Belzung, L. Segu, R. Misslin, M. C. Buhot, and R. Hen. "5-HT1B Receptor Knock Out—Behavioral Consequences." *Behavioral Brain Research* 73, no. 1–2 (1996): 305–12.

Ramsbotham, L. D., and B. Gesch. "Crime and Nourishment: Cause for a Rethink?" *Prison Service Journal* 182 (Mar. 1, 2009): 3–9.

Sen, A., N. Heredia, M. C. Senut, S. Land, K. Hollocher, X. Lu, M. O. Dereski, and D. M. Ruden. "Multigenerational Epigenetic Inheritance in Humans: DNA Methylation Changes Associated With Maternal Exposure to Lead Can Be Transmitted to the Grandchildren." *Scientific Reports* 5 (Sep. 29, 2015): 14466.

Tiihonen, J., M. R. Rautiainen, H. M. Ollila, E. Repo-Tiihonen, M. Virkkunen, A. Palotie, O. Pietilainen, et al. "Genetic Background of Extreme Violent Behavior." *Molecular Psychiatry* 20, no. 6 (June 2015): 786–92.

Torrey, E. F., J. J. Bartko, and R. H. Yolken. "*Toxoplasma gondii* and Other Risk Factors for Schizophrenia: An Update." *Schizophrenia Bulletin* 38, no. 3 (May 2012): 642–47.

Weissman, M. M., V. Warner, P. J. Wickramaratne, and D. B. Kandel. "Maternal Smoking During Pregnancy and Psychopathology in Offspring Followed to Adulthood." *Journal of the American Academy of Child and Adolescent Psychiatry* 38, no. 7 (July 1999): 892–99.

Chapter 7

Acevedo, B. P., A. Aron, H. E. Fisher, and L. L. Brown. "Neural Correlates of Long-Term Intense Romantic Love." *Social Cognitive and Affective Neuroscience* 7, no. 2 (Feb. 2012): 145–59.

Barash, D.P., and J.E. Lipton. *The Myth of Monogamy: Fidelity and Infidelity in Animals and People*. New York: W. H. Freeman, 2001.

Buston, P. M., and S. T. Emlen. "Cognitive Processes Underlying Human Mate Choice: The Relationship Between Self-Perception and Mate Preference in Western Society." *Proceedings of the National Academy of the Sciences USA* 100, no. 15 (July 22, 2003): 8805–10.

Ciani, A. C., F. Iemmola, and S. R. Blecher. "Genetic Factors Increase Fecundity in Female Maternal Relatives of Bisexual Men as in Homosexuals." *Journal of Sexual Medicine* 6, no. 2 (Feb. 2009): 449–55.

Conley, T. D., J. L. Matsick, A. C. Moors, and A. Ziegler. "Investigation of Consensually Nonmonogamous Relationships." *Perspectives on Psychological Science* 12, no. 2 (2017): 205–32.

De Dreu, C. K., L. L. Greer, G. A. Van Kleef, S. Shalvi, and M. J. Handgraaf. "Oxytocin Promotes Human Ethnocentrism." *Proceedings of the National Academy of Sciences USA* 108, no. 4 (Jan. 25, 2011): 1262–66.

Feldman, R., A. Weller, O. Zagoory-Sharon, and A. Levine. "Evidence for a Neuroendocrinological Foundation of Human Affiliation: Plasma Oxytocin Levels Across Pregnancy and the Postpartum Period Predict Mother-Infant Bonding." *Psychological Science* 18, no. 11 (Nov. 2007): 965–70.

Fillion, T. J., and E. M. Blass. "Infantile Experience With Suckling Odors Determines Adult Sexual Behavior in Male Rats." *Science* 231, no. 4739 (Feb. 14, 1986): 729–31.

Finkel, E. J., J. L. Burnette, and L. E. Scissors. "Vengefully Ever After: Destiny Beliefs, State Attachment Anxiety, and Forgiveness." *Journal of Personality and Social Psychology* 92, no. 5 (May 2007): 871–86.

Fisher, H., A. Aron, and L. L. Brown. "Romantic Love: An fMRI Study of a Neural Mechanism for Mate Choice." *Journal of Comparative Neurology* 493, no. 1 (Dec. 5, 2005): 58–62.

Fisher, Helen. *Anatomy of Love: A Natural History of Mating, Marriage, and Why We Stray.* New York: W. W. Norton & Company, 2016.

Fraccaro, P. J., B. C. Jones, J. Vukovic, F. G. Smith, C. D. Watkins, D. R. Feinberg, A. C. Little, and L. M. DeBruine. "Experimental Evidence That Women Speak in a Higher Voice Pitch to Men They Find Attractive." *Journal of Evolutionary Psychology* 9, no. 1 (2011): 57–67.

Garcia, J. R., J. MacKillop, E. L. Aller, A. M. Merriwether, D. S. Wilson, and J. K. Lum. "Associations Between Dopamine D4 Receptor Gene Variation With Both Infidelity and Sexual Promiscuity." *PLoS One* 5, no. 11 (Nov. 30, 2010): e14162.

Ghahramani, N. M., T. C. Ngun, P. Y. Chen, Y. Tian, S. Krishnan, S. Muir, L. Rubbi, et al. "The Effects of Perinatal Testosterone Exposure on the DNA Methylome of the Mouse Brain Are Late-Emerging." *Biology of Sex Differences* 5 (2014): 8.

Gobrogge, K. L., and Z. W. Wang. "Genetics of Aggression in Voles." *Advances in Genetics* 75 (2011): 121–50.

Hamer, D. H., S. Hu, V. L. Magnuson, N. Hu, and A. M. Pattatucci. "A Linkage Between DNA Markers on the X Chromosome and Male Sexual Orientation." *Science* 261, no. 5119 (July 16, 1993): 321–27.

Hanson, Joe. "The Odds of Finding Life and Love." It's Okay to Be Smart, www.youtube.com/watch?time_continue=254&v=TekbxvnvYb8.

Havlíček, J., R. Dvořáková, L. Bartoš, and J. Flegr. "Non-Advertized Does Not Mean Concealed: Body Odour Changes Across the Human Menstrual Cycle." *Ethology* 112, no. 1 (2006): 81–90.

Kimchi, T., J. Xu, and C. Dulac. "A Functional Circuit Underlying Male Sexual Behaviour in the Female Mouse Brain." *Nature* 448, no. 7157 (Aug. 30, 2007): 1009–14.

Lee, S., and N. Schwarz. "Framing Love: When It Hurts to Think We Were Made for Each Other." *Journal of Experimental Social Psychology* 54 (2014): 61–67.

LeVay, S. "A Difference in Hypothalamic Structure Between Heterosexual and Homosexual Men." *Science* 253, no. 5023 (Aug. 30, 1991): 1034–37.

Lim, M. M., Z. Wang, D. E. Olazabal, X. Ren, E. F. Terwilliger, and L. J. Young. "Enhanced Partner Preference in a Promiscuous Species by Manipulating the Expression of a Single Gene." *Nature* 429, no. 6993 (June 17, 2004): 754–57.

Marazziti, D., H. S. Akiskal, A. Rossi, and G. B. Cassano. "Alteration of the Platelet Serotonin Transporter in Romantic Love." *Psychological Medicine* 29, no. 3 (May 1999): 741–45.

Marazziti, D., H. S. Akiskal, M. Udo, M. Picchetti, S. Baroni, G. Massimetti, F. Albanese, and L. Dell'Osso. "Dimorphic Changes of Some Features of Loving Relationships During Long-Term Use of Antidepressants in Depressed Outpatients." *Journal of Affective Disorders* 166 (Sep. 2014): 151–55.

Meyer-Bahlburg, H. F., C. Dolezal, S. W. Baker, and M. I. New. "Sexual Orientation in Women With Classical or Non-Classical Congenital Adrenal Hyperplasia as a Function of Degree of Prenatal Androgen Excess." *Archives of Sexual Behavior* 37, no. 1 (Feb. 2008): 85–99.

Morran, L. T., O. G. Schmidt, I. A. Gelarden, R. C. Parrish, II, and C. M. Lively. "Running With the Red Queen: Host-Parasite Coevolution Selects for Biparental Sex." *Science* 333, no. 6039 (July 8, 2011): 216–18.

Munroe, Randall. *What If?: Serious Scientific Answers to Absurd Hypothetical Questions*. New York: Houghton Mifflin Harcourt, 2014.

Ngun, T. C., and E. Vilain. "The Biological Basis of Human Sexual Orientation: Is There a Role for Epigenetics?" *Advances in Genetics* 86 (2014): 167–84.

Nugent, B. M., C. L. Wright, A. C. Shetty, G. E. Hodes, K. M. Lenz, A. Mahurkar, S. J. Russo, S. E. Devine, and M. M. McCarthy. "Brain Feminization Requires Active Repression of Masculinization Via DNA Methylation." *Nature Neuroscience* 18, no. 5 (May 2015): 690–97.

Odendaal, J. S., and R. A. Meintjes. "Neurophysiological Correlates of Affiliative Behaviour Between Humans and Dogs." *Veterinary Journal* 165, no. 3 (May 2003): 296–301.

Paredes-Ramos, P., M. Miquel, J. Manzo, and G. A. Coria-Avila. "Juvenile Play Conditions Sexual Partner Preference in Adult Female Rats." *Physiology & Behavior* 104, no. 5 (Oct. 24, 2011): 1016–23.

Paredes, R. G., T. Tzschentke, and N. Nakach. "Lesions of the Medial Preoptic Area/Anterior Hypothalamus (MPOA/HA) Modify Partner Preference in Male Rats." *Brain Research* 813, no. 1 (Nov. 30, 1998): 1–8.

Park, D., D. Choi, J. Lee, D. S. Lim, and C. Park. "Male-Like Sexual Behavior of Female Mouse Lacking Fucose Mutarotase." *BMC Genetics* 11 (July 7, 2010): 62.

Pedersen, C. A., and A. J. Prange, Jr. "Induction of Maternal Behavior in Virgin Rats After Intracerebroventricular Administration of Oxytocin." *Proceedings of the National Academy of the Sciences USA* 76, no. 12 (Dec. 1979): 6661–65.

Ramsey, J. L., J. H. Langlois, R. A. Hoss, A. J. Rubenstein, and A. M. Griffin. "Origins of a Stereotype: Categorization of Facial Attractiveness by 6-Month-Old Infants." *Developmental Science* 7, no. 2 (Apr. 2004): 201–11.

Rhodes, G. "The Evolutionary Psychology of Facial Beauty." *Annual Review of Psychology* 57 (2006): 199–226.

Sanders, A. R., G. W. Beecham, S. Guo, K. Dawood, G. Rieger, J. A. Badner, E. S. Gershon, et al. "Genome-Wide Association Study of Male Sexual Orientation." *Scientific Reports* 7, no. 1 (Dec. 7, 2017): 16950.

Sanders, A. R., E. R. Martin, G. W. Beecham, S. Guo, K. Dawood, G. Rieger, J. A. Badner, et al. "Genome-Wide Scan Demonstrates Significant Linkage for Male Sexual Orientation." *Psychological Medicine* 45, no. 7 (May 2015): 1379–88.

Sansone, R. A., and L. A. Sansone. "Ssri-Induced Indifference." *Psychiatry (Edgmont)* 7, no. 10 (Oct. 2010): 14–18.

Scheele, D., A. Wille, K. M. Kendrick, B. Stoffel-Wagner, B. Becker, O. Gunturkun, W. Maier, and R. Hurlemann. "Oxytocin Enhances Brain Reward System Responses in Men Viewing the Face of Their Female Partner." *Proceedings of the National Academy of the Sciences USA* 110, no. 50 (Dec. 10, 2013): 20308–13.

Sharon, G., D. Segal, J. M. Ringo, A. Hefetz, I. Zilber-Rosenberg, and E. Rosenberg. "Commensal Bacteria Play a Role in Mating Preference of *Drosophila melanogaster.*" *Proceedings of the National Academy of the Sciences USA* 107, no. 46 (Nov. 16, 2010): 20051–56.

Singh, D. "Female Mate Value at a Glance: Relationship of Waist-to-Hip Ratio to Health, Fecundity and Attractiveness." *Neuro Endocrinology Letters* 23 Suppl. 4 (Dec. 2002): 81–91.

Singh, D., and D. Singh. "Shape and Significance of Feminine Beauty: An Evolutionary Perspective." *Sex Roles* 64, no. 9–10 (2011): 723–31.

Stern, K., and M. K. McClintock. "Regulation of Ovulation by Human Pheromones." *Nature* 392, no. 6672 (Mar. 12, 1998): 177–79.

Swami, V., and M. J. Tovee. "Resource Security Impacts Men's Female Breast Size Preferences." *PLoS One* 8, no. 3 (2013): e57623.

Thornhill, R., and S. W. Gangestad. "Facial Attractiveness." *Trends in Cognitive Sciences* 3, no. 12 (Dec. 1999): 452–60.

Walum, H., L. Westberg, S. Henningsson, J. M. Neiderhiser, D. Reiss, W. Igl, J. M. Ganiban, et al. "Genetic Variation in the Vasopressin Receptor 1a Gene (AVPR1A) Associates with Pair-Bonding Behavior in Humans." *Proceedings of the National Academy of the Sciences USA* 105, no. 37 (Sep. 16, 2008): 14153–56.

Wedekind, C., T. Seebeck, F. Bettens, and A. J. Paepke. "Mhc-Dependent Mate Preferences in Humans." *Proceedings: Biological Sciences* 260, no. 1359 (June 22, 1995): 245–49.

Weisman, O., O. Zagoory-Sharon, and R. Feldman. "Oxytocin Administration to Parent Enhances Infant Physiological and Behavioral Readiness for Social Engagement." *Biological Psychiatry* 72, no. 12 (Dec. 15, 2012): 982–89.

Williams, J. R., C. S. Carter, and T. Insel. "Partner Preference Development in Female Prairie Voles Is Facilitated by Mating or the Central Infusion of Oxytocin." *Annals of the New York Academy of Sciences* 652 (June 12, 1992): 487–89.

Winslow, J. T., N. Hastings, C. S. Carter, C. R. Harbaugh, and T. R. Insel. "A Role for Central Vasopressin in Pair Bonding in Monogamous Prairie Voles." *Nature* 365, no. 6446 (Oct. 7, 1993): 545–48.

Witt, D. M., and T. R. Insel. "Central Oxytocin Antagonism Decreases Female Reproductive Behavior." *Annals of the New York Academy of Sciences* 652 (June 12, 1992): 445–47.

Zeki, S. "The Neurobiology of Love." *FEBS Letters* 581, no. 14 (June 12, 2007): 2575–79.

Zuniga, A., R. J. Stevenson, M. K. Mahmut, and I. D. Stephen. "Diet Quality and the Attractiveness of Male Body Odor." *Evolution & Human Behavior* 38, no. 1 (2017): 136–43.

Chapter 8

Bellinger, D. C. "A Strategy for Comparing the Contributions of Environmental Chemicals and Other Risk Factors to Neurodevelopment of Children." *Environmental Health Perspectives* 120, no. 4 (Apr. 2012): 501–07.

Bench, S. W., H. C. Lench, J. Liew, K. Miner, and S. A. Flores. "Gender Gaps in Overestimation of Math Performance." *Sex Roles* 72, no. 11–12 (2015): 536–46.

Biergans, S. D., C. Claudianos, J. Reinhard, and C. G. Galizia. "DNA Methylation Mediates Neural Processing After Odor Learning in the Honeybee." *Scientific Reports* 7 (Feb. 27, 2017): 43635.

Brass, M., and P. Haggard. "To Do or Not to Do: The Neural Signature of Self-Control." *Journal of Neuroscience* 27, no. 34 (Aug. 22, 2007): 9141–45.

Bustin, G. M., D. N. Jones, M. Hansenne, and J. Quoidbach. "Who Does Red Bull Give Wings To? Sensation Seeking Moderates Sensitivity to Subliminal Advertisement." *Frontiers in Psychology* 6 (2015): 825.

Claro, S., D. Paunesku, and C. S. Dweck. "Growth Mindset Tempers the Effects of Poverty on Academic Achievement." *Proceedings of the National Academy of the Sciences USA* 113, no. 31 (Aug. 2, 2016): 8664–68.

Danziger, S., J. Levav, and L. Avnaim-Pesso. "Extraneous Factors in Judicial Decisions." *Proceedings of the National Academy of the Sciences USA* 108, no. 17 (Apr. 26, 2011): 6889–92.

Else-Quest, N. M., J. S. Hyde, and M. C. Linn. "Cross-National Patterns of Gender Differences in Mathematics: A Meta-Analysis." *Psychological Bulletin* 136, no. 1 (Jan. 2010): 103–27.

Fitzsimons, G. M., T. Chartrand, and G. J. Fitzsimons. "Automatic Effects of Brand Exposure on Motivated Behavior: How Apple Makes You 'Think Different.' " *Journal of Consumer Research* 35 (2008): 21–35.

Gareau, M. G., E. Wine, D. M. Rodrigues, J. H. Cho, M. T. Whary, D. J. Philpott, G. Macqueen, and P. M. Sherman. "Bacterial Infection Causes Stress-Induced Memory Dysfunction in Mice." *Gut* 60, no. 3 (Mar. 2011): 307–17.

Graff, J., and L. H. Tsai. "The Potential of HDAC Inhibitors as Cognitive Enhancers." *Annual Review of Pharmacology and Toxicology* 53 (2013): 311–30.

Hariri, A. R., T. E. Goldberg, V. S. Mattay, B. S. Kolachana, J. H. Callicott, M. F. Egan, and D. R. Weinberger. "Brain-Derived Neurotrophic Factor Val66Met Polymorphism Affects Human Memory-Related Hippocampal Activity and Predicts Memory Performance." *Journal of Neuroscience* 23, no. 17 (July 30, 2003): 6690–94.

Hart, W., and D. Albarracin. "The Effects of Chronic Achievement Motivation and Achievement Primes on the Activation of Achievement and Fun Goals." *Journal of Personality and Social Psychology* 97, no. 6 (Dec. 2009): 1129–41.

Jasarevic, E., C. L. Howerton, C. D. Howard, and T. L. Bale. "Alterations in the Vaginal Microbiome by Maternal Stress Are Associated With Metabolic Reprogramming of the Offspring Gut and Brain." *Endocrinology* 156, no. 9 (Sep. 2015): 3265–76.

Jones, M. W., M. L. Errington, P. J. French, A. Fine, T. V. Bliss, S. Garel, P. Charnay, et al. "A Requirement for the Immediate Early Gene Zif268 in the Expression of Late LTP and Long-Term Memories." *Nature Neuroscience* 4, no. 3 (Mar. 2001): 289–96.

Kaufman, G. F., and L. K. Libby. "Changing Beliefs and Behavior Through Experience-Taking." *Journal of Personality and Social Psychology* 103, no. 1 (July 2012): 1–19.

Kida, S., and T. Serita. "Functional Roles of CREB as a Positive Regulator in the Formation and Enhancement of Memory." *Brain Research Bulletin* 105 (June 2014): 17–24.

Kramer, M. S., F. Aboud, E. Mironova, I. Vanilovich, R. W. Platt, L. Matush, S. Igumnov, et al. "Breastfeeding and Child Cognitive Development: New Evidence From a Large Randomized Trial." *Archives of General Psychiatry* 65, no. 5 (May 2008): 578–84.

Krenn, B. "The Effect of Uniform Color on Judging Athletes' Aggressiveness, Fairness, and Chance of Winning." *Journal of Sport and Exercise Psychology* 37, no. 2 (Apr. 2015): 207–12.

Kruger, J., and D. Dunning. "Unskilled and Unaware of It: How Difficulties in Recognizing One's Own Incompetence Lead to Inflated Self-Assessments." *Journal of Personality and Social Psychology* 77, no. 6 (Dec. 1999): 1121–34.

Kuhn, S., D. Kugler, K. Schmalen, M. Weichenberger, C. Witt, and J. Gallinat. "The Myth of Blunted Gamers: No Evidence for Desensitization in Empathy for Pain After a Violent Video Game Intervention in a Longitudinal fMRI Study on Non-Gamers." *Neurosignals* 26, no. 1 (Jan. 31, 2018): 22–30.

Letzner, S., O. Gunturkun, and C. Beste. "How Birds Outperform Humans in Multi-Component Behavior." *Current Biology* 27, no. 18 (Sep 25 2017): R996-R98.

Libet, B., C. A. Gleason, E. W. Wright, and D. K. Pearl. "Time of Conscious Intention to Act in Relation to Onset of Cerebral Activity (Readiness-Potential). The Unconscious Initiation of a Freely Voluntary Act." *Brain* 106 (Pt. 3; Sep. 1983): 623–42.

Mackay, D. F., G. C. Smith, S. A. Cooper, R. Wood, A. King, D. N. Clark, and J. P. Pell. "Month of Conception and Learning Disabilities: A Record-Linkage Study of 801,592 Children." *American Journal of Epidemiology* 184, no. 7 (Oct. 1, 2016): 485–93.

Miller, B. L., J. Cummings, F. Mishkin, K. Boone, F. Prince, M. Ponton, and C. Cotman. "Emergence of Artistic Talent in Frontotemporal Dementia." *Neurology* 51, no. 4 (Oct. 1998): 978–82.

Murphy, S. T., and R. B. Zajonc. "Affect, Cognition, and Awareness: Affective Priming with Optimal and Suboptimal Stimulus Exposures." *Journal of Personality and Social Psychology* 64, no. 5 (May 1993): 723–39.

Robinson, G. E., and A. B. Barron. "Epigenetics and the Evolution of Instincts." *Science* 356, no. 6333 (Apr. 7, 2017): 26–27.

Rydell, R. J., A. R. McConnell, and S. L. Beilock. "Multiple Social Identities and Stereotype Threat: Imbalance, Accessibility, and Working Memory." *Journal of Personality and Social Psychology* 96, no. 5 (May 2009): 949–66.

Sniekers, S., S. Stringer, K. Watanabe, P. R. Jansen, J. R. I. Coleman, E. Krapohl, E. Taskesen, et al. "Genome-Wide Association Meta-Analysis of 78,308 Individuals Identifies New Loci and Genes Influencing Human Intelligence." *Nature Genetics* 49, no. 7 (July 2017): 1107–12.

Snyder, A. W., E. Mulcahy, J. L. Taylor, D. J. Mitchell, P. Sachdev, and S. C. Gandevia. "Savant-Like Skills Exposed in Normal People by Suppressing the Left Fronto-Temporal Lobe." *Journal of Integrative Neuroscience* 2, no. 2 (Dec. 2003): 149–58.

Soon, C. S., M. Brass, H. J. Heinze, and J. D. Haynes. "Unconscious Determinants of Free Decisions in the Human Brain." *Nature Neuroscience* 11, no. 5 (May 2008): 543–45.

Stein, D. J., T. K. Newman, J. Savitz, and R. Ramesar. "Warriors Versus Worriers: The Role of Comt Gene Variants." *CNS Spectrums* 11, no. 10 (Oct. 2006): 745–48.

Tang, Y. P., E. Shimizu, G. R. Dube, C. Rampon, G. A. Kerchner, M. Zhuo, G. Liu, and J. Z. Tsien. "Genetic Enhancement of Learning and Memory in Mice." *Nature* 401, no. 6748 (Sep. 2, 1999): 63–69.

Webster, G. D., G. R. Urland, and J. Correll. "Can Uniform Color Color Aggression? Quasi-Experimental Evidence From Professional Ice Hockey." *Social Psychological and Personality Science* 3, no. 3 (2011): 274–81.

Wimmer, M. E., L. A. Briand, B. Fant, L. A. Guercio, A. C. Arreola, H. D. Schmidt, S. Sidoli, et al. "Paternal Cocaine Taking Elicits Epigenetic Remodeling and Memory Deficits in Male Progeny." *Molecular Psychiatry* 22, no. 11 (Nov. 2017): 1641–50.

Chapter 9

Blanke, O., and S. Arzy. "The Out-of-Body Experience: Disturbed Self-Processing at the Temporo-Parietal Junction." *Neuroscientist* 11, no. 1 (Feb. 2005): 16–24.

Block, J., and J. H. Block. "Nursery School Personality and Political Orientation Two Decades Later." *Journal of Research in Personality* 40 (2006): 734–49.

Borjigin, J., U. Lee, T. Liu, D. Pal, S. Huff, D. Klarr, J. Sloboda, et al. "Surge of Neurophysiological Coherence and Connectivity in the Dying Brain." *Proceedings of the National Academy of the Sciences USA* 110, no. 35 (Aug. 27, 2013): 14432–37.

Carney, D. R., J. T. Jost, S. D. Gosling, and J. Potter. "The Secret Lives of Liberals and Conservatives: Personality Profiles, Interaction Styles, and the Things They Leave Behind." *Politicial Psychology* 29, no. 6 (2008): 807–40.

Caspar, E. A., J. F. Christensen, A. Cleeremans, and P. Haggard. "Coercion Changes the Sense of Agency in the Human Brain." *Current Biology* 26, no. 5 (Mar. 7, 2016): 585–92.

Chawla, L. S., S. Akst, C. Junker, B. Jacobs, and M. G. Seneff. "Surges of Electroencephalogram Activity at the Time of Death: A Case Series." *Journal of Palliative Medicine* 12, no. 12 (Dec. 2009): 1095–100.

Eidelman, S., C. S. Crandall, J. A. Goodman, and J. C. Blanchar. "Low-Effort Thought Promotes Political Conservatism." *Personality and Social Psychology Bulletin* 38, no. 6 (June 2012): 808–20.

Haider-Markel, D. P., and M. R. Joslyn. " 'Nanny' State Politics: Causal Attributions About Obesity and Support for Regulation." *American Politics Research* 46, no. 2 (2017): 199–216.

Holstege, G., J. R. Georgiadis, A. M. Paans, L. C. Meiners, F. H. van der Graaf, and A. A. Reinders. "Brain Activation During Human Male Ejaculation." *Journal of Neuroscience* 23, no. 27 (Oct. 8 2003): 9185–93.

Horne, Z., D. Powell, J. E. Hummel, and K. J. Holyoak. "Countering Antivaccination Attitudes." *Proceedings of the National Academy of the Sciences USA* 112, no. 33 (Aug. 18, 2015): 10321–24.

Janoff-Bulman, R. "To Provide or Protect: Motivational Bases of Political Liberalism and Conservatism." *Psychological Inquiry* 20, no. 2–3 (2009): 120–28.

Kanai, R., T. Feilden, C. Firth, and G. Rees. "Political Orientations Are Correlated With Brain Structure in Young Adults." *Current Biology* 21, no. 8 (Apr. 26, 2011): 677–80.

Kaplan, J. T., S. I. Gimbel, and S. Harris. "Neural Correlates of Maintaining One's Political Beliefs in the Face of Counterevidence." *Scientific Reports* 6 (Dec. 23, 2016): 39589.

Konnikova, Maria. "The Real Lesson of the Standford Prison Experiment." www.newyorker.com/science/maria-konnikova/the-real-lesson-of-the-stanford-prison-experiment.

Levine, M., A. Prosser, D. Evans, and S. Reicher. "Identity and Emergency Intervention: How Social Group Membership and Inclusiveness of Group Boundaries Shape Helping Behavior." *Personality and Social Psychology Bulletin* 31, no. 4 (Apr. 2005): 443–53.

Musolino, Julien. *The Soul Fallacy: What Science Shows We Gain From Letting Go of Our Soul Beliefs*. Amherst, NY: Prometheus Books, 2015.

Oxley, D. R., K. B. Smith, J. R. Alford, M. V. Hibbing, J. L. Miller, M. Scalora, P. K. Hatemi, and J. R. Hibbing. "Political Attitudes Vary With Physiological Traits." *Science* 321, no. 5896 (Sep. 19, 2008): 1667–70.

Parnia, S., K. Spearpoint, G. de Vos, P. Fenwick, D. Goldberg, J. Yang, J. Zhu, et al. "Aware-Awareness During Resuscitation-a Prospective Study." *Resuscitation* 85, no. 12 (Dec. 2014): 1799–805.

Paul, G. S. "Cross-National Correlations of Quantifiable Societal Health With Popular Religiosity and Secularism in the Prosperous Democracies." *Journal of Religion and Society* 7 (2005).

Pinker, S. "The Brain: The Mystery of Consciousness." TIME, http://content.time.com/time/magazine/article/0,9171,1580394-1,00.html.

Sample, Ian. "Stephen Hawking: 'There Is No Heaven; It's a Fairy Story.' " *The*

Guardian, www.theguardian.com/science/2011/may/15/stephen-hawking-interview-there-is-no-heaven.

Settle, J. E., C. T. Dawes, N. A. Christakis, and J. H. Fowler. "Friendships Moderate an Association Between a Dopamine Gene Variant and Political Ideology." *Journal of Politics* 72, no. 4 (2010): 1189–98.

Sharot, Tali. *The Influential Mind: What the Brain Reveals About Our Power to Change Others*. New York: Henry Holt and Co., 2017.

Sunstein, C. R., S. Bobadilla-Suarez, S. Lazzaro, and T. Sharot. "How People Update Beliefs About Climate Change: Good News and Bad News." *Social Science Research Network* (2016): https://ssrn.com/abstract=2821919.

University, Emory. "Emory Study Lights Up the Political Brain." www.emory.edu/news/Releases/PoliticalBrain1138113163.html.

Westen, D., P. S. Blagov, K. Harenski, C. Kilts, and S. Hamann. "Neural Bases of Motivated Reasoning: An fMRI Study of Emotional Constraints on Partisan Political Judgment in the 2004 U.S. Presidential Election." *Journal of Cognitive Neuroscience* 18, no. 11 (Nov. 2006): 1947–58.

Chapter 10

Anderson, S. C., J. F. Cryan, and T. Dinan. *The Psychobiotic Revolution: Mood, Food, and the New Science of the Gut-Brain Connection*. Washington, D.C.: National Geographic, 2017.

Armstrong, D., and M. Ma. "Researcher Controls Colleague's Motions in 1st Human Brain-to-Brain Interface." UW News, Aug. 27, 2013.

Benito E., C. Kerimoglu, B. Ramachandran, T. Pena-Centeno, G. Jain, R. M. Stilling, M. R. Islam, V. Capece, Q. Zhou, D. Edbauer, C. Dean, and A. Fischer. "RNA-Dependent Intergenerational Inheritance of Enhanced Synaptic Plasticity After Environmental Enrichment." *Cell Reports* 23, no. 2 (Apr. 10, 2018): 546–54.

Benmerzouga, I., L. A. Checkley, M. T. Ferdig, G. Arrizabalaga, R. C. Wek, and W. J. Sullivan, Jr. "Guanabenz Repurposed as an Antiparasitic With Activity Against Acute and Latent Toxoplasmosis." *Antimicrobial Agents and Chemotherapy* 59, no. 11 (Nov. 2015): 6939–45.

Berger, T. W., R. E. Hampson, D. Song, A. Goonawardena, V. Z. Marmarelis, and S. A. Deadwyler. "A Cortical Neural Prosthesis for Restoring and Enhancing Memory." *Journal of Neural Engineering* 8, no. 4 (Aug. 2011): 046017.

Bieszczad, K. M., K. Bechay, J. R. Rusche, V. Jacques, S. Kudugunti, W. Miao, N. M. Weinberger, J. L. McGaugh, and M. A. Wood. "Histone Deacetylase Inhibition Via RGFP966 Releases the Brakes on Sensory Cortical Plasticity and the Specificity of Memory Formation." *Journal of Neuroscience* 35, no. 38 (Sep. 23, 2015): 13124–32.

Cavazzana-Calvo, M., S. Hacein-Bey, G. de Saint Basile, F. Gross, E. Yvon, P. Nusbaum, F. Selz, et al. "Gene Therapy of Human Severe Combined Immunodeficiency (SCID)-X1 Disease." *Science* 288, no. 5466 (Apr. 28, 2000): 669–72.

Chueh, A. C., J. W. Tse, L. Togel, and J. M. Mariadason. "Mechanisms of Histone Deacetylase Inhibitor-Regulated Gene Expression in Cancer Cells." *Antioxidants & Redox Signaling* 23, no. 1 (July 1, 2015): 66–84.

Cott, Emma. "Prosthetic Limbs, Controlled by Thought." *New York Times,* www.nytimes.com/2015/05/21/technology/a-bionic-approach-to-prosthetics-controlled-by-thought.html.

Desbonnet, L., L. Garrett, G. Clarke, B. Kiely, J. F. Cryan, and T. G. Dinan. "Effects of the Probiotic *Bifidobacterium infantis* in the Maternal Separation Model of Depression." *Neuroscience* 170, no. 4 (Nov. 10, 2010): 1179–88.

Eichler, F., C. Duncan, P. L. Musolino, P. J. Orchard, S. De Oliveira, A. J. Thrasher, M. Armant, et al. "Hematopoietic Stem-Cell Gene Therapy for Cerebral Adrenoleukodystrophy." *New England Journal of Medicine* 377, no. 17 (Oct. 26, 2017): 1630–38.

Guan, J. S., S. J. Haggarty, E. Giacometti, J. H. Dannenberg, N. Joseph, J. Gao, T. J. Nieland, et al. "HDAC2 Negatively Regulates Memory Formation and Synaptic Plasticity." *Nature* 459, no. 7243 (May 7, 2009): 55–60.

Hemmings, S. M. J., S. Malan-Muller, L. L. van den Heuvel, B. A. Demmitt, M. A. Stanislawski, D. G. Smith, A. D. Bohr, et al. "The Microbiome in Posttraumatic Stress Disorder and Trauma-Exposed Controls: An Exploratory Study." *Psychosomatic Medicine* 79, no. 8 (Oct. 2017): 936–46.

Hochberg, L. R., M. D. Serruya, G. M. Friehs, J. A. Mukand, M. Saleh, A. H. Caplan, A. Branner, et al. "Neuronal Ensemble Control of Prosthetic Devices by a Human With Tetraplegia." *Nature* 442, no. 7099 (July 13, 2006): 164–71.

Hsiao, E. Y., S. W. McBride, S. Hsien, G. Sharon, E. R. Hyde, T. McCue, J. A. Codelli, et al. "Microbiota Modulate Behavioral and Physiological Abnormalities Associated With Neurodevelopmental Disorders." *Cell* 155, no. 7 (Dec. 19, 2013): 1451–63.

Jacka, F. N., A. O'Neil, R. Opie, C. Itsiopoulos, S. Cotton, M. Mohebbi, D. Castle, et al. "A Randomised Controlled Trial of Dietary Improvement for Adults With Major Depression (the 'Smiles' Trial)." *BMC Medicine* 15, no. 1 (Jan 30, 2017): 23.

Kaliman, P., M. J. Alvarez-Lopez, M. Cosin-Tomas, M. A. Rosenkranz, A. Lutz, and R. J. Davidson. "Rapid Changes in Histone Deacetylases and Inflammatory Gene Expression in Expert Meditators." *Psychoneuroendocrinology* 40 (Feb. 2014): 96–107.

Kilgore, M., C. A. Miller, D. M. Fass, K. M. Hennig, S. J. Haggarty, J. D. Sweatt, and G. Rumbaugh. "Inhibitors of Class 1 Histone Deacetylases Reverse Contextual Memory Deficits in a Mouse Model of Alzheimer's Disease." *Neuropsychopharmacology* 35, no. 4 (Mar. 2010): 870–80.

Kindt M., M. Soeter, and B. Vervliet. "Beyond Extinction: Erasing Human Fear Responses and Preventing the Return of Fear." *Nature Neuroscience* 12, no. 3 (Mar. 2009): 256–58.

Liang, P., Y. Xu, X. Zhang, C. Ding, R. Huang, Z. Zhang, J. Lv, et al. "CRISPR/Cas9-Mediated Gene Editing in Human Tripronuclear Zygotes." *Protein Cell* 6, no. 5 (May 2015): 363–72.

Lindholm, M. E., S. Giacomello, B. Werne Solnestam, H. Fischer, M. Huss, S. Kjellqvist, and C. J. Sundberg. "The Impact of Endurance Training on Human Skeletal Muscle Memory, Global Isoform Expression and Novel Transcripts." *PLoS Genetics* 12, no. 9 (Sep. 2016): e1006294.

Messaoudi, M., R. Lalonde, N. Violle, H. Javelot, D. Desor, A. Nejdi, J. F. Bisson, et al. "Assessment of Psychotropic-Like Properties of a Probiotic Formulation *(Lactobacillus helveticus* R0052 and *Bifidobacterium longum* R0175) in Rats and Human Subjects." *British Journal of Nutrition* 105, no. 5 (Mar. 2011): 755–64.

Olds, D. L., J. Eckenrode, C. R. Henderson, Jr., H. Kitzman, J. Powers, R. Cole, K. Sidora, et al. "Long-Term Effects of Home Visitation on Maternal Life Course and Child Abuse and Neglect. Fifteen-Year Follow-up of a Randomized Trial." *JAMA* 278, no. 8 (Aug. 27, 1997): 637–43.

Seckel, Scott. "Asu Researcher Creates System to Control Robots With the Brain." ASU Now, asunow.asu.edu/20160710-discoveries-asu-researcher-creates-system-control-robots-brain.

Silverman, L. R. "Targeting Hypomethylation of DNA to Achieve Cellular Differentiation in Myelodysplastic Syndromes (MDS)." *Oncologist* 6 Suppl. 5 (2001): 8–14.

Singh, R. K., H. W. Chang, D. Yan, K. M. Lee, D. Ucmak, K. Wong, M. Abrouk, et al. "Influence of Diet on the Gut Microbiome and Implications for Human Health." *Journal of Translational Medicine* 15, no. 1 (Apr. 8, 2017): 73.

Sleiman, S. F., J. Henry, R. Al-Haddad, L. El Hayek, E. Abou Haidar, T. Stringer, D. Ulja, et al. "Exercise Promotes the Expression of Brain Derived Neurotrophic Factor (BDNF) through the Action of the Ketone Body Beta-Hydroxybutyrate." *Elife* 5 (June 2, 2016).

Weaver, I. C., N. Cervoni, F. A. Champagne, A. C. D'Alessio, S. Sharma, J. R. Seckl, S. Dymov, M. Szyf, and M. J. Meaney. "Epigenetic Programming by Maternal Behavior." *Nature Neuroscience* 7, no. 8 (Aug. 2004): 847–54.

INDEX

TK

Index

TK

TK

Index

TK

TK

Index

TK

ABOUT THE AUTHOR

William "Bill" Sullivan, Jr., obtained his Ph.D. in cell and molecular biology at the University of Pennsylvania and is now Showalter Professor at the Indiana University School of Medicine in Indianapolis, where he studies microbiology and genetics. He has published nearly 100 scientific papers and holds a patent for antiparasitic drug discovery. Known among students as the "funny T-shirt guy," Bill has earned multiple teaching awards for his lectures. Bill has been featured in *Scientific American;* on CNN Health, IFLScience!, and *The Naked Scientists;* and at Gen Con, and more. Through his writing, presentations, and interviews, he strives to make science accessible and entertaining for all.

You can find him at authorbillsullivan.com or on Twitter @wjsullivan.